OIL SPI
FROM
TANK VESSEL
LIGHTERING

Committee on Oil Spill Risks from Tank Vessel Lightering
Marine Board
Commission on Engineering and Technical Systems
National Research Council

NATIONAL ACADEMY PRESS
Washington, D.C.   1998

NATIONAL ACADEMY PRESS • 2101 Constitution Avenue, NW • Washington, D.C. 20418

NOTICE: The project that is the subject of this report was approved by the Governing Board of the National Research Council, whose members are drawn from the councils of the National Academy of Sciences, the National Academy of Engineering, and the Institute of Medicine. The members of the committee responsible for the report were chosen for their special competencies and with regard for appropriate balance.

The National Academy of Sciences is a private, nonprofit, self-perpetuating society of distinguished scholars engaged in scientific and engineering research, dedicated to the furtherance of science and technology and to their use for the general welfare. Upon the authority of the charter granted to it by the Congress in 1863, the Academy has a mandate that requires it to advise the federal government on scientific and technical matters. Dr. Bruce M. Alberts is president of the National Academy of Sciences.

The National Academy of Engineering was established in 1964, under the charter of the National Academy of Sciences, as a parallel organization of outstanding engineers. It is autonomous in its administration and in the selection of its members, sharing with the National Academy of Sciences the responsibility for advising the federal government. The National Academy of Engineering also sponsors engineering programs aimed at meeting national needs, encourages education and research, and recognizes the superior achievements of engineers. Dr. William A. Wulf is president of the National Academy of Engineering.

The Institute of Medicine was established in 1970 by the National Academy of Sciences to secure the services of eminent members of appropriate professions in the examination of policy matters pertaining to the health of the public. The Institute acts under the responsibility given to the National Academy of Sciences by its congressional charter to be an adviser to the federal government and, upon its own initiative, to identify issues of medical care, research, and education. Dr. Kenneth I. Shine is president of the Institute of Medicine.

The National Research Council was organized by the National Academy of Sciences in 1916 to associate the broad community of science and technology with the Academy's purposes of furthering knowledge and advising the federal government. Functioning in accordance with general policies determined by the Academy, the Council has become the principal operating agency of both the National Academy of Sciences and the National Academy of Engineering in providing services to the government, the public, and the scientific and engineering communities. The Council is administered jointly by both Academies and the Institute of Medicine. Dr. Bruce M. Alberts and Dr. William A. Wulf are chairman and vice chairman, respectively, of the National Research Council.

This study was supported by Contract DTMA91-94-G-00003 between the National Academy of Sciences and the Maritime Administration of the U.S. Department of Transportation. Any opinions, findings, conclusions, or recommendations expressed in this publication are those of the author(s) and do not necessarily reflect the views of the organizations or agencies that provided support for the project.

Library of Congress Catalog Card Number 98-75577

International Standard Book Number 0-309-06190-3

Limited copies are available from: Marine Board, Commission on Engineering and Technical Systems, National Research Council, 2101 Constitution Avenue, N.W., Washington, D.C. 20418.

Additional copies of this report are available from National Academy Press, 2101 Constitution Avenue, N.W., Lockbox 285, Washington, D.C. 20055; (800) 624-6242 or (202) 334-3313 (in the Washington metropolitan area); Internet, http://www.nap.edu

Cover Photo Credit: OMI Petrolink.

Printed in the United States of America

COMMITTEE ON OIL SPILL RISKS FROM
TANK VESSEL LIGHTERING

DON E. KASH, *chair*, George Mason University, Fairfax, Virginia
TRICIA CLARK, Oil Spill Division of Texas General Land Office, Austin
ALFRED COLE, Chevron Shipping Company, Pascagoula, Mississippi
EDWARD C. CROSS, Plimsoll Shipping, Inc., Houston, Texas
DUANE H. LAIBLE, Glosten Associates, Inc., Seattle, Washington
J. BRADFORD MOONEY, JR., consultant, Arlington, Virginia
STEPHEN D. RICKS, Clean Bay, Inc., Concord, California
EDWIN J. ROLAND, Bona Shipping (U.S.), Inc., Houston, Texas
RICHARD J. STEADY, Maritrans Operating Partners, L.P., Philadelphia,
    Pennsylvania
JOHN B. TORGAN, Save the Bay, Providence, Rhode Island
W.M. von ZHAREN, Texas A&M University, Galveston

Liaisons of Sponsoring Agencies

STEPHEN L. KANTZ, U.S. Coast Guard, Washington, D.C.
ZELVIN LEVINE, U.S. Maritime Administration, Washington, D.C.

Marine Board Liaison

STEVEN T. SCALZO, Foss Maritime Company, Seattle, Washington

Marine Board Staff

PETER A. JOHNSON, Acting Director
THERESA M. FISHER, Administrative Assistant
LAURA OST, Consultant

iii

iv

Preface

Lightering first emerged as a common practice in U.S. waters, particularly the Gulf of Mexico, almost 30 years ago. Historically, there has been little reason to question the safety of lightering, despite the apparent risks involved in transferring cargo between two vessels that are under way, drifting, or anchored at sea. However, public concerns about oil spills have drawn congressional attention to lightering in recent years. Concerns have also been raised by some observers of the maritime industry about the safety implications of projected increases in waterborne U.S. oil imports and certain lightering-related provisions of the Oil Pollution Act of 1990 (P.L. 101-380).

The Coast Guard Authorization Act of 1996 (P.L. 104-324) requires the U.S. Coast Guard (USCG) to coordinate with the Marine Board of the National Research Council to conduct studies on the risks of oil spills from lightering operations off U.S. coasts. Accordingly, an 11-member committee was convened under the auspices of the Marine Board. The committee was asked to accomplish the following tasks:

- investigate the frequency and risks of accidents from lightering operations
- assess the existing regulatory and management framework
- recommend measures that could reduce the risks of oil spills

The committee was constituted to include members with expertise in tanker and lightering operations, risk assessment, regulatory and management approaches to accident prevention, oil spill accident analysis, environmental protection, international rules and standards, and USCG prevention and enforcement practices. The principle guiding the constitution of the committee and its work, consistent with NRC policy, was not to exclude members with potential biases if they had expertise vital to the study but to seek balance and fair treatment of the subject. The biographies of committee members are provided in Appendix A.

The committee met four times over a one-year period, and subgroups of the committee held one additional meeting and visited three lightering operations in the Gulf of Mexico and one in Delaware Bay (a complete list of meetings and presenters is provided in Appendix B). The committee also reviewed previous studies of lightering safety and collected data on accidents and spills from the USCG, industry groups and individual companies, and state agencies. The data are summarized in the report, and additional details are provided in appendices C, D, and E. Additional background information collected for this study included relevant letters (Appendix F), regulations (Appendix G), and industry safety guidelines and checklists (Appendix H).

The study focused narrowly on the risks of oil spills from lightering, which was defined as the transfer at sea of crude oil or petroleum products from one vessel to another. The committee attempted to identify and assess only those spills that could be directly attributed to the lightering operation rather than spills that may have occurred coincidentally during the approach, transfer, or post-transfer phases of lightering. Bunkering, automatic cargo shutoff valves, formal risk assessment, and a comparison of lightering and alternative methods of oil delivery were beyond the scope of the study.

The committee wishes to acknowledge the contributions of the more than two dozen representatives of industry and government who made presentations during meetings and shared additional background information in other contexts. All of this information was crucial to the study process, especially in light of deficiencies in existing accident databases and the decentralization of the lightering industry.

The committee wishes to acknowledge USCG liaison Lieutenant Commander Stephen L. Kantz, who provided background on the regulatory regime and USCG data on lightering-related accidents and spills. In addition, the committee wishes to thank the persons and organizations listed in Appendix B for their extra efforts and valuable contributions to the study. These include: the Coast Guard Captains of the Port and Marine Safety Office personnel in Houston, Galveston, Philadelphia, Long Island Sound, San Francisco, and San Diego; members of the Industry Taskforce on Offshore Lightering; the firms, SeaRiver Maritime, Skaugen PetroTrans, Inc., Chevron Shipping Company, and Maritrans, who provided special data and hosted visits by committee members; and Dr. Zelvin Levine, the Maritime Administration liaison to the study.

This report has been reviewed in draft form by individuals chosen for their diverse perspective and technical expertise, in accordance with procedures approved by the NRC's Report Review Committee. The purpose of this independent review is to provide candid and critical comments that will assist the institution in making the published report as sound as possible and to ensure that the report meets institutional standards for objectivity, evidence, and responsiveness to the study charge. The review comments and draft manuscript remain confidential to protect the integrity of the deliberative process. We wish to thank the following individuals for their participation in the review of this report:

Gary L. Borman, University of Wisconsin
Dennis L. Bryant, Haight, Garden, Poor and Havens
John W. Farrington, Woods Hole Oceanographic Institution
Michael J. Herz, Consultant
R. Keith Michel, Herbert Engineering
Ronald P. Nordgren, Rice University
Malcolm L. Spaulding, University of Rhode Island
Emmett G. Ward, Texas A&M University

While the individuals listed above have provided constructive comments and suggestions, it must be emphasized that responsibility for the final content of this report rests entirely with the authoring committee and the institution.

Contents

Figures, Tables, and Boxes

FIGURES

TABLES

BOXES

OIL SPILL RISKS
FROM
TANK VESSEL
LIGHTERING

Executive Summary

The safety record of lightering (the transfer of petroleum cargo at sea from a large tanker to smaller ones) has been excellent in U.S. waters in recent years, as evidenced by the very low rate of spillage of oil both in absolute terms and compared with all other tanker-related accidental spills. The lightering safety record is likely to be maintained or even improved in the future as overall quality improvements in the shipping industry are implemented. Risks can be reduced even further through measures that enhance sound lightering standards and practices, support cooperative industry efforts to maintain safety, and increase the availability of essential information to shipping companies and mariners. Only continued vigilance and attention to safety initiatives can avert serious accidents involving tankers carrying large volumes of oil.

LIGHTERING ACTIVITY IN THE UNITED STATES

Lightering is an effective and cost-efficient method of delivering foreign crude oil to U.S. refineries and importing petroleum products. Lightering is necessary because very large tankers, which are often used to move cargo from the Arabian Gulf and other distant sources of oil, are too wide and too deep to enter most U.S. ports. Transferring part or all of the cargo to smaller vessels for delivery to terminals is less expensive than moving all of the cargo the entire distance in a larger number of smaller vessels.

Lightering safety became a topic of national interest several years ago because of public concerns about oil spills in general. The Coast Guard Authorization Act of 1996 requires that the U.S. Coast Guard (USCG) coordinate with the Marine Board of the National Research Council (NRC) to conduct studies on the risks of oil spills from lightering off the U.S. coasts. Accordingly, an 11-member

1

committee was assembled by the NRC, under the auspices of the Marine Board, to evaluate current lightering practices and trends, analyze the safety record, assess the regulatory and standards-setting framework, analyze accident prevention and risk reduction measures, and recommend technical and institutional improvements. The highlights of the one-year study and the committee's 16 recommendations are summarized below.

More than 25 percent of the 7.5 million barrels of crude oil imported into the United States each day is lightered. Small amounts of refined products are also lightered.[1] Approximately 95 percent of offshore lightering (i.e., outside the territorial sea, which generally extends three miles off the U.S. coastline), by volume, takes place in the Gulf of Mexico, according to government data. Additional offshore lightering takes place off Long Island, near the New Jersey and Virginia capes, off San Diego in California, and near the Bahamas. More than two-thirds of inshore lightering (i.e., within the territorial sea), by volume, takes place on the East Coast, primarily in the Delaware Bay and River and Long Island Sound. The rest of the inshore lightering takes place on the West Coast, in San Francisco Bay. The committee's estimates of the volume of oil involved in inshore lightering, combined with government data on offshore lightering, provide the most complete picture of U.S. lightering activity available to date. Although the projected increase of U.S. oil imports may lead to an increase in lightering, the committee expects that increases in the near term will be small and that current lightering patterns and volumes will remain fairly steady.

In the following sections, the vessel from which the cargo is removed is referred to as the ship to be lightered (STBL), and the receiving vessel is referred to as the service vessel. The STBLs and service vessels may either be owned by an oil company or chartered on a long-term basis or for a specific voyage. The STBLs are typically large tankers. A number of U.S. companies are engaged solely in the lightering business and operate service vessels. Service vessels may be all-purpose tankers, tankers equipped specifically for lightering, integrated tug-barge units equipped specifically for lightering, or standard all-purpose tug-barge units.

LIGHTERING SAFETY AND SPILL RECORD

Even though an immense amount of information is available on maritime accidents in general, it is difficult to collect reliable data on the history of oil spills related to lightering in U.S. waters. The deficiencies in the data include inconsistent reporting and ambiguous information on the underlying causes of accidents. For this study, the accident data, which varied greatly in detail and reliability, were collected from various sources, including the USCG, state agencies, shipping companies and organizations, and other private sources. The

---

[1]Offshore lightering currently accounts for about 80 percent and inshore lightering about 20 percent of the grand total of lightering volumes.

committee attempted to identify and assess only those spills that were attributable directly to mishaps during lightering operations.

The combined data showed an excellent safety record, which was verified by substantial anecdotal evidence and numerous interviews, as well as the personal experience of committee members. The USCG data for 1984-1996 indicate that few spills occurred during lightering on U.S. coasts, and the average spill volume was only 26 barrels (1,095 gallons). Recurring causes of spills that appear to be directly related to lightering include valve failures, tank overflows, and hose ruptures. The committee collected additional data from the USCG, industry, and state agencies for 1993-1997. During that time, no spills were reported on the east or west coasts, and only seven spills (accounting for less than 0.003 percent of the total volume lightered) were reported in the Gulf of Mexico. Only one spill was substantial, resulting from a collision in 1995 near Galveston, Texas; in that case, more than 850 barrels (35,700 gallons) of fuel oil were spilled. The cause of the accident was attributed to both limited communications and procedural errors, according to the USCG.

EFFECTIVENESS OF REGULATIONS, STANDARDS, AND PRACTICES

Various controls have been imposed on lightering (and tanker operators in general) by international agreements and U.S. laws and regulations. The USCG oversees lightering operations outside port areas through six general mechanisms: vessel design requirements, operational procedures, personnel qualifications, oil spill contingency planning and equipment requirements, vessel inspection, and monitoring. Three separate sets of regulations have been promulgated by the USCG regarding lightering activities. One set applies to lightering in inshore waters. For this purpose, inshore waters means all waters inside of 12 nautical miles from the coast, including all internal waters (i.e., lakes, bays, sounds, and rivers). The second set of regulations applies to lightering in all offshore waters, except for designated lightering zones. Offshore, for this purpose, means between 12 and 200 miles off the coast. The third, and most comprehensive, set of regulations applies in designated lightering zones that are more than 60 miles off the coast. The Coast Guard does not regulate lightering in foreign waters or outside the U.S. Exclusive Economic Zone (EEZ). Technically, lightering in offshore waters is subject to regulation by the Coast Guard only when the cargo is bound for a U.S. port. As a practical matter, though, all oil lightered in U.S. waters is bound for the United States. Under the comprehensive national lightering regulations, four areas are designated lightering zones (offshore) in the Gulf of Mexico.

In general, lightering is performed with the local USCG captain of the port (COTP) exercising regulatory authority. The regulatory regime for lightering is widely regarded as adequate, with one notable exception. Industry representatives informed the committee that vessels sometimes have to maneuver excessively or separate prematurely to comply with a legal provision that requires certain vessels to remain within designated lightering zones in the Gulf of Mexico except in emergencies.

Industry guidelines for lightering have been established by at least two industry groups, and most individual companies have developed their own internal guidelines. A set of comprehensive minimum standards for offshore lightering, now in its third edition, has been developed by the Oil Companies International Marine Forum (OCIMF), an international group of vessel owners and charterers. The guidelines contain advice on lightering procedures and arrangements, as well as specifications for mooring, fenders, and cargo transfer hoses. The shipping industry relies heavily on the OCIMF guidelines, although anecdotal evidence and the observations of committee members suggest that the guidelines are not applied uniformly. In the United States, a supplement to the OCIMF guidelines was developed by the Industry Taskforce on Offshore Lightering (ITOL), a highly effective cooperative organization that promotes industry self-policing and, in partnership with the USCG, continuous improvement in lightering in the Gulf of Mexico.

The OCIMF guidelines are also widely used for U.S. inshore lightering. General standards for inland shipping have been established by the American Waterways Operators (AWO), but no separate lightering standards have been established for inland trade despite its unique characteristics, such as the extensive use of barges and the frequent transport of specialized refined products.

The OCIMF and ITOL lightering guidelines leave little room for improvement, with the exception of a few details concerning the characteristics of the safest vessel designs and equipment. Procedural issues requiring special emphasis include measures for ensuring that key vessel personnel are fluent in English and that the safest methods are used for gauging cargo following a lightering operation.

The committee also paid particular attention to human factors in its investigation of spill risks and noted that the training and certification of all personnel engaged in lightering are critical to safe operations. The international maritime industry has recently adopted a new convention on training and certification for seafarers that represents a milestone for enhancing the skills and competency of shipboard personnel in the international fleet as a whole. In addition, the committee found that the lightering industry's training programs and qualification requirements for mooring masters and other expert advisors are based on high standards and rigorous oversight. These programs and requirements should continue to serve as a sound model in the future.

GAPS IN AVAILABLE INFORMATION

The safety of lightering depends heavily on the information available to shipping and lightering companies and vessel personnel. The committee identified three major gaps in the availability, appropriateness, and accuracy of the information that is readily available for lightering purposes.

First, one of the major variables in lightering is the condition and performance of the STBL and its operators. Although information about the condition of a vessel can be obtained from the owner/operator, lightering companies do not currently have direct access to the more comprehensive, vetted information in the SIRE[2] database, an oil industry initiative that facilitates the sharing of data on vessel condition among OCIMF members. Second, the current systems for providing marine weather forecasts in the United States do not fully meet lightering needs. Reported problems include the inappropriate location of weather buoys, a lack of real-time information, and delays in repairs of buoys and data links. Third, to avoid anchoring on and damaging underwater oil pipelines in the Gulf of Mexico, the operators of STBLs and service vessels need accurate data on pipeline locations. Federal agencies do not currently collect and publish these data on a timely basis.

RECOMMENDATIONS

The sound safety record of lightering in U.S. waters, combined with the strong preference of the maritime community for self-regulation, suggests that most measures aimed at reducing the risk of spillage would be best performed by the industry. Accordingly, 10 of the committee's 16 recommendations are addressed to shipping and lightering companies and organizations. The remaining six recommendations, which require national leadership or intersect with agency missions, are addressed to the USCG and other federal agencies.

Recommendations for Shipping Companies and Organizations

Recommendation 1. Industry organizations, such as the American Waterways Operators, or cooperative organizations modeled on the Industry Taskforce on Offshore Lightering should develop standards and guidelines for inshore lightering operations. This document could either supplement or incorporate appropriate sections of the Oil Companies International Marine Forum guidelines for offshore operations.

Recommendation 2. Chartering organizations should screen all prospective ships to be lightered to determine whether they meet Oil Companies International Marine Forum standards for vessels, equipment, and crews and should not charter vessels that do not meet these standards. As a supplementary measure to

---

[2]This database of technical information, which concerns the condition and operational procedures of tankers in the world fleet, is known as the Ship Inspection Report (SIRE) program. SIRE is maintained by OCIMF, the tanker industry organization.

determine whether this self-policing process is effective, the U.S. Coast Guard should monitor the process and call for periodic reports when appropriate and needed.

Recommendation 3. The Oil Companies International Marine Forum should consider making limited revisions to its Ship Inspection Report regulations to give lightering companies access to information on the condition of vessels.

Recommendation 4. To promote the adequate rigging of secondary fenders, the Oil Companies International Marine Forum should emphasize (e.g., in the next edition of its lightering guidelines) the need for vertical and flat surfaces as high as possible along vessel sides above the load waterline, with the maximum amount of vertical sides consistent with design requirements. In addition, mounting points, leads, and lifting equipment for secondary fenders should be positioned and sized for optimum effectiveness, and leads and securing facilities should be provided for primary fenders to ensure maximum coverage.

Recommendation 5. The Oil Companies International Marine Forum should focus on the need for vessels to have enough full-sized mooring bitts and enclosed chocks to secure the two vessels together with a minimum of four lines forward and aft. A minimum of one full-sized mooring bitt and enclosed chock should be provided within 35 meters forward and aft of the manifold. All mooring lines should be secured by winches.

Recommendation 6. The Oil Companies International Marine Forum should focus on the need for vessels that are capable of slow steaming for extended periods of time (within the limited operating range of modern diesel engines) with fine control of engine revolutions to enable safe maneuvering during mooring and unmooring operations.

Recommendation 7. The Oil Companies International Marine Forum should recommend limited operating parameters for modern double-hull tankers used as ships to be lightered to accommodate excessive freeboard (up to about 85 feet) when the cargo tanks are empty, a condition that can degrade the integrity of the mooring between the ship to be lightered and the service vessel. At the same time, the International Maritime Organization should consider modifying MARPOL. Annex I, Regulation 13 or classifying lightering as an "exceptional case" to permit greater ballasting of these vessels when transferring oil to a service vessel.

Recommendation 8. Mooring lines should be fitted with synthetic tails and fenders designed specifically for lightering operations. Lightering operators should also adhere carefully to existing standards and guidelines with regard to the inspection and testing of hoses.

Recommendation 9. Before initiating cargo transfer operations, the mooring master (or equivalent person in charge) aboard the service vessel should determine whether the key individuals on the ship to be lightered are fluent in English and can understand the lightering plans and respond to commands. If necessary, an individual (reporting to the lightering master or other official in charge of lightering) who is both fluent in English and knowledgeable about lightering should be put aboard the ship to be lightered prior to the transfer of cargo.

Recommendation 10. To limit the time that vessels are alongside each other in a seaway and avoid delays in departure under adverse or marginal weather conditions, the Industry Taskforce on Offshore Lightering should suggest (e.g., in the next edition of its offshore lightering guidelines) that the mooring master and vessel master dispense with the inspector's gauging (at least on the service vessel) until the vessel is in port. If cargo quantities must be ascertained offshore, gauging should be limited to the ship to be lightered and should be done after the service vessel has departed. The cargo measurements for the service vessel could be telexed to the ship to be lightered.

Recommendations for the U.S. Coast Guard and Other Federal Agencies

Recommendation 11. The U.S. Coast Guard should encourage the lightering industry on the east and west coasts to adopt or adapt the Industry Taskforce on Offshore Lightering model as part of their program to promote problem solving, interaction, and cooperation to enhance safety and environmental protection. Cooperative arrangements could be initiated through existing mechanisms, such as the American Waterways Operators/U.S. Coast Guard Safety Partnership.

Recommendation 12. The U.S. Coast Guard, in consultation with the lightering industry, should work with the National Weather Service and the U.S. Navy to select locations for weather buoys and to tailor weather data and forecasts to support operations in both designated lightering zones and traditional lightering areas. The National Weather Service should take on this task as a priority to improve the delivery of weather information for offshore operations.

Recommendation 13. The U.S. Coast Guard captain of the port should be given the authority, based on a case-by-case review of individual requests and circumstances, to allow vessels to leave designated lightering zones for safety reasons while still engaged in lightering.

Recommendation 14. The U.S. Coast Guard should develop, or hire a private contractor to develop, an accurate, comprehensive computer database on maritime oil spills that can be searched and sorted by pertinent variables, including the causes of accidents.

Recommendation 15. The Minerals Management Service (of the U.S. Department of the Interior) and the Office of Pipeline Safety (of the U.S. Department of Transportation) should develop and implement a plan to collect and compile accurate data on the location of pipelines in the Gulf of Mexico and make the information available to the operators of vessels that engage in lightering. Priority should be placed on data collection in designated lightering zones and traditional lightering areas, and the data should be verified and updated on a regular basis.

Recommendation 16. To ensure safe anchorages amid the increasing oil and gas exploration activity, the U.S. Coast Guard should seek authority to designate "pipeline-free areas" where new pipelines cannot be laid.

1

Introduction

The idea of transferring crude oil or petroleum products between two vessels that are under way, anchored, or drifting on the open ocean may seem risky. And yet, according to shipping companies and maritime accident statistics, this common practice—known as lightering—is safe, as long as certain conditions are met. More than 25 percent of the more than 7.5 million barrels of crude oil imported to the United States each day is lightered, or transferred from one vessel to another, before delivery to port.[1] A comparable proportion of the 6.4 million barrels of oil per day produced domestically is carried by water and lightered (DOE, 1998). Few vessel accidents or spills that are directly attributable to lightering operations have ever occurred in U.S. waters. Nevertheless, public concerns about oil spills makes it important to maintain vigilance over lightering activities.

Almost 30 years have passed since lightering first became a routine practice in the U.S. Gulf of Mexico and Delaware Bay, driven by increases in crude oil imports and in the size of ships (NRC, 1997). Over the past few decades, strong economic incentives have led to the use of very large tankers for the long hauls from the Persian Gulf and Africa. Because these ships are too deep and too wide to approach or enter most U.S. ports safely, the oil cargo is transferred to smaller vessels that deliver it to refineries. As the cost of the domestic oil supply has risen and its availability has declined, East Coast refiners have become dependent

---

[1]Approximately 25 percent of imported oil is lightered offshore (outside the U.S. territorial sea), according to data for 1994 provided by the Maritime Administration Office of Statistical Analysis, Washington, D.C. Additional imported oil is lightered closer to shore, but the government does not maintain specific records on this activity. The committee's estimates are provided later in this chapter.

A TYPICAL LIGHTERING OPERATION

The lightering process begins with a service vessel making its approach to an STBL .

After the vessels are moored together, hoses are passed from the service vessel to the STBL and connected to the cargo oil piping.

At the final approach, the two ships are alongside each other, protected by fenders.

In this case, the service vessel is equipped with a cargo crane to connect the transfer hoses. Photo Credit: Chevron Shipping Co.

After the transfer is complete, the two vessels are unmoored, and the service vessel departs to deliver the oil.

Two vessels are moored together during the transfer of cargo from the STBL to the service vessel.

on imported crude oil. The frequency of lightering has led both the shipping industry and the U.S. Coast Guard (USCG) to take a number of steps to ensure the safety of this practice.

Although these safety initiatives have been effective so far, several factors suggest that a review of the risks and practices of lightering is warranted. First, lightering activity levels or patterns could change in the near future, although no dramatic changes are expected. Oil imports into the U.S. Gulf of Mexico have been steadily increasing, and, if more imports are shipped across long distances, then lightering operations may become more frequent. Product and crude oil offshore lightering activity has also increased in recent years along the East Coast, and there is a possibility of continued small increases.[2] Second, lightering might be encouraged by the Oil Pollution Act of 1990 (OPA 90; P.L. 101-380), which prohibits certain tank vessels from approaching U.S. shores (NRC, 1997). Operators of these vessels will have to lighter their cargo instead. In any case, the lightering industry will have to maintain, or even improve, its existing high performance levels by continuously evolving the standards established at various levels by industry and national and international regulatory bodies.

The U.S. Congress has also expressed concern about lightering safety. The present study was conducted to satisfy the requirements of the Coast Guard Authorization Act of 1996 (P.L. 104-324), Section 903, which requires an assessment of the risks of oil spills attributable to lightering operations off the U.S. coasts. This study evaluates statistics on oil spills and existing spill-prevention measures, examines current activities and future trends in offshore lightering, assesses the regulatory framework and standards and operations for lightering, and recommends measures to reduce the risk of oil spills and minimize environmental damage. The study does not address port and terminal activities, bunkering, or cargo spills that are not directly related to lightering. Nor does it assess the relative risks associated with alternative methods of transferring cargo, such as using deepwater ports or offshore pipelines.

LIGHTERING AT A GLANCE

For the purposes of this report, "lightering" is defined as the mooring of two vessels to transfer petroleum cargo,[3] excluding bunkers, between the ship to be lightered (STBL) and the receiving vessel (the service vessel) for the purpose of either taking cargo from, or adding cargo to, the STBL (see Box 1-1). Most often the service vessel takes on cargo for delivery to a shore terminal.

Lightering typically takes place either 12 or more miles offshore or at deepwater anchorages and other sheltered locations inshore. (The term "inshore"

---

[2]Inshore crude oil lightering on the East Coast is not expected to increase.

[3]Although many types of cargo can be lightered, this report focuses on crude oil and petroleum products to fulfill the statutory mandate of evaluating the risks of oil spills.

BOX 1-1 DEFINITIONS

Lightering: Lightering is the mooring of two vessels for the purpose of transferring petroleum cargo, excluding bunkers, from the ship to be lightered (STBL) to a service vessel. The process can be divided into three phases: the approach phase, the transfer phase, and the post-transfer phase.

Vessels Involved in Lightering: The STBL is generally either too large or has too deep a draft to enter the port or facility where the cargo is to be delivered. The service vessel may be of several types: all-purpose tankers, tankers specifically equipped for lightering, integrated tug-barge units specifically equipped for lightering, and standard all-purpose tug-barge units.

Lightering Processes: In the open ocean, the two vessels are moored and unmoored either while both are under way or while the STBL is at anchor. The process is completed without the assistance of tugboats or other vessels. During inshore lightering, the STBL is always at anchor and in a semi-protected or protected area.

is defined here as inside the outer boundary of the contiguous zone, which the United States extends 12 miles from shore.) Lightering is usually necessary because the STBL is too large, or has too deep a draft,[4] to enter the harbor or approach the terminal where the cargo is to be picked up or delivered. The service vessel may be an all-purpose tanker, a tanker equipped specifically for lightering, an integrated tug-barge unit equipped specifically for lightering, or a standard all-purpose tug-barge unit.

Lightering operations in the open ocean may span a geographical area of many miles and may take several hours to a week or more, depending on the number of discharges and the volume of cargo discharged. Lightering takes place when both vessels are under way or drifting or when the STBL is at anchor. The process encompasses three phases: the approach phase, the transfer phase, and post-transfer phase. The approach phase begins when the two vessels are approximately three miles apart. Once they are moored together and cargo transfer begins, one "lift"—a discharge from an STBL to a service vessel—can take 10 to 24 hours to complete. Very large STBLs can require as many as eight lifts to transfer all cargo; thus, they may remain within a lightering area for as long as 20

---

[4]The term "draft" refers to the depth of a vessel below the waterline. Large ships have deeper drafts than small vessels, and all vessels have deeper drafts (i.e., sit deeper in the water) when fully loaded with cargo than when empty or partially loaded.

days to discharge their cargo completely. The process proceeds more quickly when service vessels are dedicated lightering vessels, which are permanently outfitted with the required equipment.

Inshore lightering usually takes place at a dedicated, deepwater anchorage in a sheltered location, such as Delaware Bay, San Francisco Bay, or Long Island Sound. The STBL is always at anchor, with the service vessel (usually a tug-barge unit) maneuvering alongside, resulting in a very limited geographical scope of operations. The process entails the same three phases as in offshore lightering, but the transfer often takes less time—usually 8 to 15 hours, depending on the size and configuration of the service vessel and the type of cargo. A typical inshore lightering operation entails no more than three lifts, which lightens the STBL enough to allow it to continue to a terminal, where the remaining cargo can be offloaded.

Lightering is usually done for economic reasons. Because of economies of scale, it is more economical to move oil in large tankers over the greatest distance possible and then, near the destination, transfer it to a smaller vessels than to move the same amount of oil the entire distance in six or seven smaller tankers. Shipping oil from the Arabian Gulf directly to a Gulf of Mexico port in service-sized vessels costs 70 percent more than moving the same amount of oil by lightering (von Zharen, 1994). Another economic reason for lightering is to avoid dead freight charges associated with "light loading" of a vessel bound for a port with a restricted depth. Cost is not the only reason for lightering, however. Lightering sometimes takes place between two vessels of the same size, not because cargo needs to be moved to another ship to reach port, but because the transfer is specified in the terms of the contract between cargo traders.

Approximately 95 percent of the offshore lightering in U.S. waters takes place in the Gulf of Mexico, where restricted water depths keep large ships from entering most ports to deliver oil to refineries. Offshore lightering also takes place off Long Island, near the New Jersey and Virginia capes, off San Diego in California, and near the Bahamas. Inshore lightering takes place in the Delaware Bay and River, Long Island Sound, New York Harbor, and San Francisco Bay (see Figure 1-1).

The largest tankers have only a few alternatives to lightering.[5] In southern California and Hawaii, ships can stop at offshore moorings and use their own pumps to transfer oil to shore, and in the Gulf of Mexico they can use the only U.S. deepwater port, the Louisiana Offshore Oil Port (LOOP), which is 18 miles south of Grand Isle, Louisiana, in approximately 115 feet of water. Between 10 and 15 percent of U.S. oil imports are brought in through the LOOP (NRC, 1997),

---

[5]The committee did not analyze alternatives to lightering, which would have been beyond the scope of this report. The committee felt however, that neither economics nor safety factors nor any business trends would force dramatic changes in current practices. However, if future trends indicate that major changes in lightering activities are likely, some site-specific alternatives could be analyzed.

FIGURE 1-1 U.S. lightering operations.

which has a capacity of approximately 1.4 million barrels per day. However, economics and the demands of the refineries supplied through the LOOP pipeline connections have limited use of the port to far below its original design capacity. The amount of imported crude oil currently passing through the LOOP is approximately 900,000 barrels per day, which is close to the maximum level achieved to date and represents a marked increase since the early 1990s (personal communication from Thomas P. James, general counsel, LOOP, Inc., April 7, 1998).

Lightering activities in the United States are described in more detail in Chapter 2. This report does not deal with foreign operations, but it is important to note that lightering is a standard link in the worldwide petroleum supply chain that is used in many parts of the world, especially in the Far East (see Box 1-2).

SAFETY RECORD

Lightering involves a series of operations, including the approach maneuver, berthing, mooring, hose connection, cargo transfer, hose disconnection, and unmooring. Lightering spills can occur for a variety of reasons, such as a ruptured hose, a tank overflow, or a collision, many of which can also occur when a vessel is unloading at a dock. The risks unique to lightering are associated with vessels coming close together (seafarers are generally trained to keep vessels apart); the breakaway procedure; severe weather; and problems with fenders, hoses, and other equipment. The greatest risk in lightering may be from human error, which

BOX 1-2 Lightering Worldwide

Lightering is common throughout the world. Generally two lightering methods are used: traditional side-by-side lightering, which is typical in the United States, and tandem lightering, in which floating storage units (mostly converted tankers) are permanently moored and connected to oil production and exporting facilities.

Traditional lightering is used in Argentina, Venezuela, West Africa, East Africa, Fujairah, Korforkhan, Gulf of Kutch, Rangoon, Singapore, Sumatra, Thailand, Hong Kong, Taiwan, South Korea, Gibraltar, Malta, and the United Kingdom.

Tandem lightering (associated with production facilities) is used in Iran, Oman, Australia, the Irish Sea, and Vietnam. As oil is produced, it is pumped from wells to offshore units and stored until the quantity is sufficient to export. At that point, a vessel is moored behind the unit; a floating hose is used to connect the two vessels. The operation requires special boats to handle the lines and hoses and sometimes a tugboat to keep the two vessels from coming into contact with each other. Tandem lightering is typically a continuous, permanent operation and is invariably backed by very large companies.

has been implicated as a cause of 80 percent of maritime accidents in general (NRC, 1994; USCG, 1995; von Zharen, 1994).

The committee also took note of some recent oil spills in Rhode Island, which have heightened public and congressional concerns about spills in general. Although these spills did not occur during lightering and were not in any way attributable to lightering, they caused environmental groups to question other tanker transportation practices, including lightering, especially practices about which public information was not generally available. Thus, lightering is cause for concern among certain environmentalists, and Senator John Chafee of Rhode Island sponsored legislation, which was ultimately passed by the U.S. Congress, that called for the present study. Yet even in Rhode Island, there is general agreement among decision makers that the safety record of local lightering is good. Nationally, the shipping industry stresses the benefits of lightering rather than the risks. For example, INTERTANKO, an organization that represents 250 independent tanker operators constituting much of the world fleet, contends that lightering is not only safe but also offers environmental benefits because it keeps large tankers away from shorelines and congested areas (personal communication from Jonathan Benner, INTERTANKO, August 5, 1997). The advantages and disadvantages of lightering in comparison to alternatives, such as delivery to deepwater ports, have been examined (e.g., USCG, 1993) and are not evaluated in the present study.

Worldwide, oil spills from all causes have been declining in both numbers and volume since the early 1980s (Etkin, 1997). Most spills are small and result from routine operations, such as cargo loading and discharging. Major collisions and groundings cause much larger spills, and a few very large spills are responsible for most of the oil spilled (NRC, 1991; Etkin, 1997). The patterns are similar in U.S. waters, where the number of spills has been declining since the 1980s. This trend was accelerated by OPA 90, which imposed structural requirements, safety and training mandates, and substantial economic penalties for spills. The reduction in spills can be attributed in large part to economic liabilities in the provisions of OPA 90 and the growing awareness of these liabilities (NRC, 1997). Meanwhile, the volume of oil spilled in U.S. waters dropped dramatically in the early 1990s and has remained very low compared to the levels of a decade or two ago. The extent of the recent improvement is evident in Figure 1-2.

Reliable data that can be used to assess the safety of lightering in U.S. waters are difficult to gather. An immense amount of information is available on maritime accidents, but identifying patterns and comparing data sets is problematic. The USCG's casualty database, recently renamed the Marine Investigation Module (MIN-MOD), is difficult to analyze, in large part because information on specific casualties is collected by local offices and is sometimes incomplete or is logged in inconsistent formats. Some spills that occur before or during lightering are not directly caused by, or related to, the lightering operation. For example, the tanker Mega Borg caught fire and spilled 3.9 million gallons of oil off the

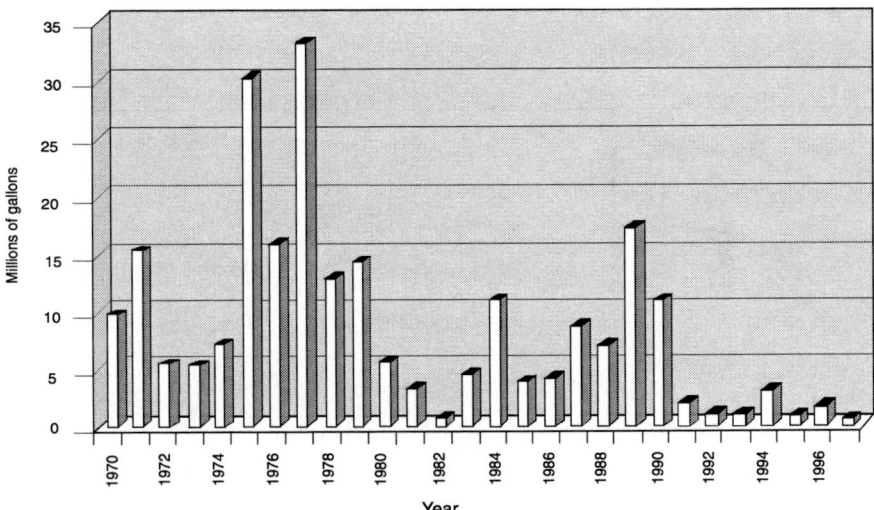

FIGURE 1-2 Oil spills of more than 10,000 gallons into U.S. maritime waters, 1970 to 1997. Source: Oil Spill Intelligence Report.

Texas coast in 1990 during a lightering operation, but the fire was not directly related to lightering. The committee considered that an explosion-and-fire accident like the one in the pump room of the Mega Borg could just as easily have happened if the ship had been unloading in port, in which case the risk to life and environmental damage could have been even greater.

The USCG provided the committee with MIN-MOD data on spills that occurred during lightering operations in U.S. waters from 1984 to 1996. The data have many limitations, most significantly that the causes of spills are either missing or are ambiguous. Despite these limitations, the committee made several general observations. First, few spills occurred during lightering operations in that time period (see Figure 1-3). Second, the average spill volume was only 26 barrels (1,095 gallons); a spill of less than 50 barrels is generally considered small. Recurring causes of spills that appear to be directly related to lightering include valve failures, tank overflows, and hose ruptures (see Figure 1-4 and Figure 1-5). The complete data are provided in Appendix C.

A formal USCG analysis of its own and other federal and private records provides additional insight. Between 1986 and 1990, 15 lightering-related accidents, resulting in total spillage of 45 barrels, were reported in the Gulf of Mexico (USCG, 1993). For the 4,391 offshore transfers in the Gulf of Mexico in those years, an average of 3.4 accidents occurred per 1,000 transfers. The average spill volume was only three barrels (126 gallons). The USCG Marine Safety Office at Galveston reported only one small spill attributable to lightering during the early

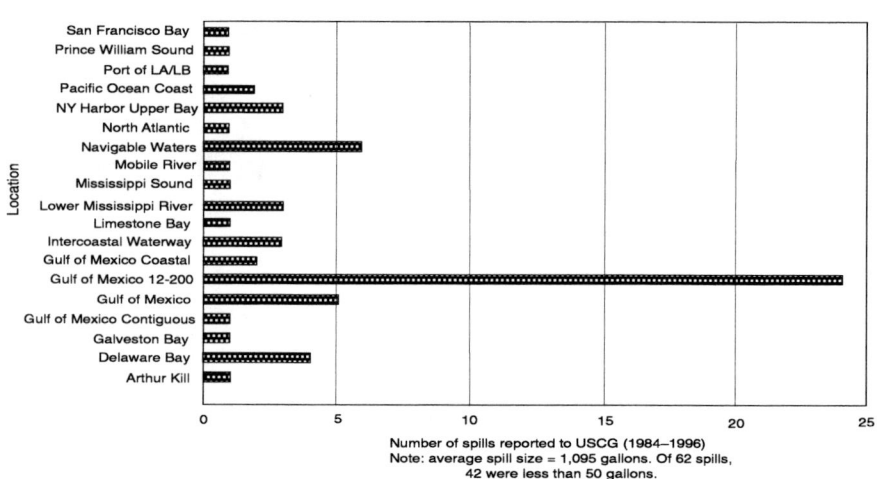

FIGURE 1-3 Location of lightering incidents.

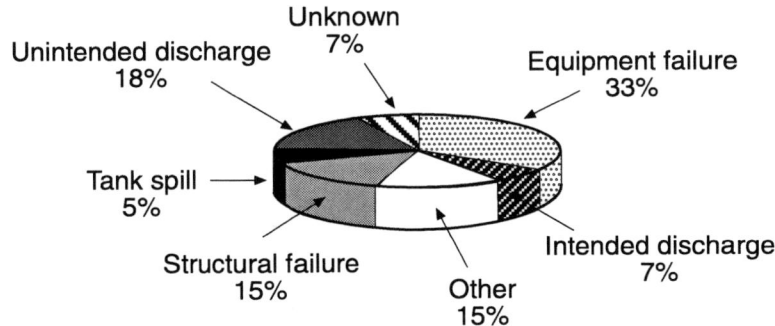

FIGURE 1-4 Lightering spills by primary cause, 1984 to 1996.

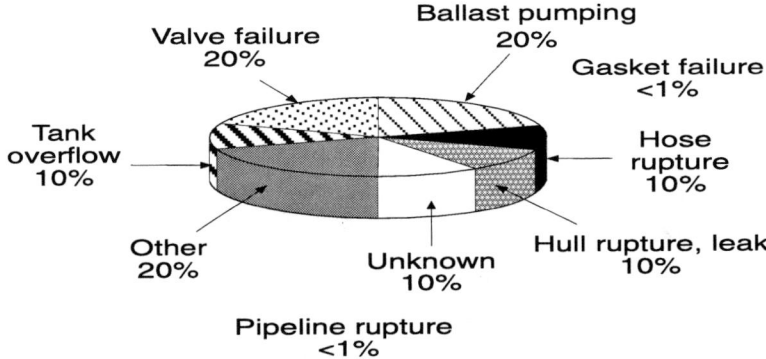

FIGURE 1-5 Lightering spills by secondary cause, 1984 to 1996.

1990s (USCG, 1993). The cause of that spill was a ruptured transfer hose. In general, the lightering safety record in the Gulf of Mexico appears to be excellent. Although the USCG data on lightering-related spills have some limitations, and some people claim that small spills offshore are less likely to be detected or reported than spills in port, the committee found no evidence of unreported lightering-related spills.[6]

Private analyses of maritime accidents (e.g., Etkin, 1997) and archives, such as those maintained by Lloyd's Register of Shipping, can be useful for some broader studies. These data are typically sorted, computerized, and subjected to

---

[6]The committee's investigation included several meetings and workshops with representatives of industry, regulatory agencies, and environmental groups (See Appendix B for a complete list). Additional resources included correspondence with shipping and lightering companies and industry organizations and the extensive experience and personal contacts of committee members.

quality assurance and quality control measures, so the records are likely to be easier to search and more reliable than the USCG database. However, private databases can also be costly to access and typically concentrate on large spills worldwide, which limits their utility for studies focusing on small spills in U.S. waters. The risks of lightering in U.S. waters have been addressed previously by independent investigators (e.g., NRC, 1991; Wilson, Gillette & Co., 1993; von Zharen, 1994), none of which found evidence of major accidents, catastrophic spills, or patterns of serious safety problems directly related to lightering.

For the present study, accident statistics for a five-year period were collected from a variety of sources, including state agencies, shipping companies, and the Industry Taskforce on Offshore Lightering (ITOL). The committee informally checked industry data with local Coast Guard offices and/or state agencies. Data from these sources were necessary because the study focused on 1993 to 1997, which is too recent to be included in some databases, and because the committee needed detailed information for all three U.S. coasts, for both the open ocean and inshore waters. The data, which are summarized in the following sections, confirm the good safety record reflected in earlier studies.

Gulf of Mexico

According to ITOL data for five lightering companies[7] operating in the Gulf of Mexico, out of approximately 5,000 lifts[8] or 2.5 billion barrels lightered between 1993 and 1997, 802[9] barrels were spilled in six accidents (see Table 1-1). Two spills occurred on deck, one from an expansion joint and one from a valve as a result of human error. Four spills went into the water: two were caused by overfilling tanks, one by an emergency breakaway, and one by a collision. The collision occurred in 1995 in the Galveston area, when the service vessel Skaubay was approaching the tanker Berge Banker to take off some of its cargo. Although no cargo was lost in the accident, a fuel tank was ruptured on the Berge Banker, and more than 850 barrels (more than 35,700 gallons) of fuel oil were spilled into the Gulf of Mexico.[10] The USCG investigation concluded that the accident was caused by limited communications about expected maneuvers (see Box 1-3). This was the only substantial spill related to lightering since 1984, when the USCG began collecting data. One oil company reported an additional spill of one barrel in 1993, when an officer on the STBL opened the manifold valves too early, and

---

[7]The five companies are Skaugen PetroTrans, Inc., American Eagle Tankers, Inc., and OMI Petrolink Corp., Aramco, and SeaRiver Maritime.

[8]A "lift" is one complete transfer operation offloading oil from an STBL to a service vessel.

[9]Most of the oil was spilled in a single accident in 1995 (the Berge Banker).

[10] Initially, the spill estimated at approximately 800 barrels, the figure reported to the committee by ITOL in 1997. The official USCG investigation, released in 1998, revised the total spill estimate to more than 850 barrels, which is the figure that appears in the text of the present report.

These vessels are preparing to begin a lightering operation in the Gulf of Mexico.

BOX 1-3 A Case of "Limited Communication"

On February 5, 1995, at 9:40 a.m. local time, the service vessel Skaubay collided with the STBL Berge Banker in the Galveston lightering area during the approach phase of a lightering operation. The following account of the incident is based on the USCG marine casualty investigation report.

The Berge Banker, a large single-hull tanker, was loaded in Saudi Arabia. The Skaubay, a small tanker, left Port Neches, Texas, on February 4, bound for the lightering site. It was expected to be the second vessel to lighter the Berge Banker.

The weather was clear, winds were from the west at 10 to 15 knots, and seas were at one to three feet. The Skaubay was traveling at 10 knots as it approached the STBL. Using a hand-held radio, the mooring master aboard the Skaubay instructed the Berge Banker to maintain a course of 270 degrees and come up to a speed of 5.5 knots. When he determined the distance between the vessels to be one mile, he ordered the Skaubay to turn to starboard. He "gave this order without knowledge of the Berge Banker's current speed . . . [he and the service vessel's captain then] realized that the Berge Banker was not going as fast as they thought it was . . . [and the mooring master] ordered the Skaubay to reduce starboard rudder angle and engine speed. However, they did not communicate the risk of collision to the Berge Banker."

The captain of the Berge Banker "was surprised at the Skaubay's course change but did not use his radar [or] radio equipment to confirm the Skaubay's intention. Based on his visual observations, he did not believe that there was a risk of collision until a few seconds before impact."

The mooring master then "ordered the Skaubay to operate astern propulsion and its rudder amidships. The captain of the Berge Banker ordered his ship to "turn hard to port, sounded the ship's internal emergency alarm, and stopped his ship. Neither ship sounded the danger signal."

When the vessels collided, the Berge Banker's fuel tank ruptured, spilling more than 850 barrels (35,700 gallons) of No. 6 fuel oil into the Gulf of Mexico. The Skaubay sustained a 25-foot gash on its port bow, whereas the STBL sustained damage to a cargo tank and bunker tank.

The Skaubay remained on scene for several days to serve as a helicopter landing platform for the pollution response operation, then sailed to Texas City for repairs. The Berge Banker, after completing the discharge of its remaining cargo, was repaired temporarily while at anchor in the lightering area. Permanent repairs were scheduled in Saudi Arabia.

BOX 1-3 *continued*

Because both vessels were Norwegian flag and the collision occurred outside the U.S. territorial sea, the government of Norway conducted the inquiry, in which the USCG participated. The collision was attributed to "limited communication between the vessels." Contributing factors included the failure of the Skaubay captain and mooring master to use all available means to determine whether there was a risk of collision when they turned their vessel toward the STBL and their failure to sound any maneuvering signal. The report also cited the failure of the Berge Banker's captain to sound a danger signal when the Skaubay turned toward his vessel.

The USCG accident report recommended that the commandant establish specific requirements regarding how a service vessel should approach an STBL. This recommendation was acted on in the final version of the lightering zone regulations and incorporated into the most recent version of the lightering guidelines issued by OCIMF (1997). (USCG, 1998)

TABLE 1-1 Lightering Incidents in the Gulf of Mexico, 1993 to 1997[a]

Incident Type	Number	Notes (including causes and volume of spill)
Vessel touching	16	13 vessel to vessel 3 workboat to vessel
Emergency separation	1	result: spill on deck
Spill on deck	2	1 expansion joint (2 gallons) 1 valve, human error (2 barrels)
Spills to environment	4	1 collision (800+ barrels)[b] 2 tank overfilling (1 barrel) 1 breakaway (1 barrel)
Injuries	3	not life threatening
Near misses	5	2 fender failures 2 engine malfunctions 1 generator fire
Other	1	1 STBL struck by workboat

[a]These are the only incidents directly attributable to lightering activities. Some additional small spills may have occurred during lightering as a result of vessel leaks or other causes that were not directly related to the lightering operation. The incidents occurred out of approximately 5,000 lifts (3 per day on average) and 2.5 billion barrels of oil transferred.

[b]This incident was the Berge Banker collision (see Box 1-3). The total spillage was later determined to be in excess of 850 barrels.

Source: Industry Taskforce on Offshore Lightering

Two vessels are lightering in Delaware Bay. Photo credit: John McGrail.

residue spilled across the deck and into the Gulf of Mexico. The committee could not determine whether this spill was or was not included in the database in Appendix C, but it was a minor spill and would not change the overall numbers substantially.

ITOL reported 16 vessel-to-vessel contacts between 1993 and 1997. STBLs came into contact with service vessels on 13 different occasions—six times during mooring, and seven times during unmooring. An additional three contacts involved workboats touching ships. Also during the same time period, five near-misses and one collision of a workboat and an STBL (with no spillage) were reported in the Gulf of Mexico. Two of these incidents were attributed to fender depressions, two to aborted moorings resulting from engine malfunctions, one to a generator fire, and one to the collision of an STBL and a workboat not involved in the lightering operations. ITOL describes the overall lightering safety record as "exemplary," noting that all incidents were followed up by the companies involved with lessons learned and risk-reduction measures. The data confirm the patterns described by the USCG (1993).

East Coast

No spills directly related to lightering were reported on the East Coast during the time period under study, according to the limited data obtained by the committee for Long Island Sound and Delaware Bay (see Appendix D). The MIN-MOD records indicate that two spills occurred along the East Coast during lightering operations between 1993 and 1997, but the causes were not recorded.

More recent USCG reports indicate that several spills occurred during, but were not directly attributed to, lightering operations in Delaware Bay. In September 1997, the tanker Mystras was engaged in a lightering operation when it spilled an estimated 20,000 gallons of oil into Delaware Bay (8,000 gallons were recovered). The cause was the failure of a valve isolating the ballast system from the cargo system. Also in late 1997, the tanker Alandia Bay spilled less than 100 gallons of oil into Delaware Bay in several separate discharges. The vessel was engaged in lightering, but the spill was traced to a heat exchanger for a vacuum pump. In 1996, the tanker Anitra spilled 500 to 800 gallons of oil into Delaware Bay when it entered the bay to engage in lightering. This spill was the result of an obstruction that prevented closure of a valve, allowing cargo to reach the ballast sea chest lines.[11]

West Coast

No lightering-related spills on the West Coast were reported for 1993 to 1997 by the California State Lands Commission, Washington Department of Ecology, or major oil companies (see Appendix E). Five gallons were spilled from a barge during one lightering operation in Long Beach, but the spill was attributed to a hull fracture, which was located by divers. The USCG MIN-MOD records reveal no spills during lightering operations along this coast since the 1980s.

Overall U.S. Lightering Spill Record for 1993 to 1997

To provide a context for evaluating lightering-related spills, the committee collected data on the total volume of cargo lightered during the same time period. Data on offshore lightering volumes only were obtained from the Maritime Administration, which does not maintain records on inshore lightering. The committee collected information from a variety of sources to try to fill in this gap. The results, shown in Table 1-2, probably underestimate the volume lightered inshore because no data could be obtained for some areas. Nevertheless, this exercise provided a more complete picture of lightering in U.S. waters than has previously been available. More than two-thirds of inshore lightering, a fairly substantial volume, takes place on the East Coast.

Table 1-2 also summarizes the combined data on spills, which show a strong safety record. Only seven spills were reported, all in the very busy Gulf of

---

[11]A "sea chest" is an integral structural box with a grill opening to the sea built on the inside of a vessel's hull. It is connected to multiple pipes (thus minimizing openings in the hull) designed for either intake or discharge of seawater from various internal systems, such as ballast water piping.

TABLE 1-2 Spills Attributed Directly to Lightering off U.S. Coasts, 1993 to 1997

Area	Barrels Lightered Offshore[d]	Barrels Lightered Inshore[e]	Total Spills	Barrels Spilled	Apparent Causes
East Coast[a]	77.5 million (2.7%)	501 million (68%)	0	0	no spills
West Coast[b]	23.5 million (0.8%)	237 million (32%)	0	0	no spills
Gulf of Mexico[c]	2.9 billion (96.5%)	0	7	> 850 (or > 35,700 gallons)	1 equipment failure 5 human error 1 unknown

[a] Information on spills for the East Coast (Delaware Bay and Long Island Sound only) was obtained from the U.S. Coast Guard. No data could be obtained for the small amount of lightering activity in other East Coast areas.

[b] Information on spills for the West Coast was obtained from the California State Lands Commission, Washington Department of Ecology, British Petroleum, Chevron, and Exxon.

[c] Information on spills for the Gulf of Mexico was obtained from the Industry Taskforce on Offshore Lightering and Chevron. One collision accounted for almost the entire volume spilled.

[d] Offshore lightering totals were estimated by the committee for the entire five-year period based on data for one year obtained from the Maritime Administration (1994).

[e] Inshore lightering totals were estimated by the committee based on data obtained from the U.S. Coast Guard for Long Island Sound and Delaware Bay on the East Coast and San Francisco Bay on the West Coast. Almost all of the East Coast inshore lightering took place in Delaware Bay. About 1 million barrels were lightered in Long Island Sound in 1997; no data could be obtained for prior years. The San Francisco figure is an estimate based on data for 1995 to 1997.

Mexico, and only one was substantial. Even more impressive, less than .003 percent of the total volume of oil lightered in the Gulf of Mexico was spilled. When the volume spilled in lightering-related incidents is compared to the overall U.S. spillage (shown previously in Figure 1-2), lightering accounted for approximately 0.5 percent of the spillage resulting from substantial incidents—those in which more than 238 barrels, or 10,000 gallons, were spilled—from 1993 to 1997.

SUMMARY

The committee's assessment confirms the results of previous studies of lightering safety that very few spills are related directly to lightering and, with rare exceptions, they are small spills. Only one substantial spill—the Berge Banker—can be attributed directly to lightering in U.S. waters from 1993 to 1997. Several other large spills occurred during lightering but were not attributed directly to the lightering operation. The committee reviewed these incidents and concluded that they probably would have occurred even if the vessels had been engaged in other activities.

Given the absence of a comprehensive, reliable database on spills directly attributable to lightering in U.S. waters, the committee was forced to collect information of varying degrees of completeness and reliability from various sources. The impression of a solid overall safety record conveyed by the resulting data was confirmed by the experience of individual committee members and by presentations and other comments by representatives of industry, government agencies, and environmental groups. The USCG's current approach to data collection and maintenance is not conducive to studies of this type.

REFERENCES

DOE (U.S. Department of Energy). 1998. Petroleum Supply Monthly. May 1998, with Data For March 1998. Energy Information Administration.

Etkin, D.S. 1997. Oil Spills from Vessels (1960–1995): An International Historical Perspective. Arlington, Mass.: Cutter Information Corp.

NRC (National Research Council). 1991. Tanker Spills: Prevention by Design. Washington, D.C.: National Academy Press.

NRC. 1994. Minding the Helm: Marine Navigation and Piloting. Washington, D.C.: National Academy Press.

NRC. 1997. Double-Hull Tanker Legislation: An Assessment of the Oil Pollution Act of 1990. Washington, D.C.: National Academy Press.

USCG (U.S. Coast Guard). 1993. Deepwater Port Study. Washington, D.C.: USCG Office of Marine Safety, Security, and Environmental Protection.

USCG. 1995. Prevention Through People Quality Action Team Report. Washington, D.C.: U.S. Department of Transportation.

USCG. 1998. Marine Casualty Investigation Report, Case MC95002001, 05 Feb 95. Washington, D.C.: U.S. Department of Transportation.

Wilson, Gillette & Co. 1993. The Evaluation of Past and Future Crude Oil Lightering Operations in the U.S. Gulf Coast. Unpublished internal report by Wilson, Gillette & Co., Arlington, Virginia.

von Zharen, W.M. 1994. Risk Evaluation of Ship-to-Ship Oil Transfer: An Assessment of Lightering as a Predictably Sound Environmental Risk: Inherent Relevant Concerns and Operational Safeguards. Galveston, Tex.: Maritime and Environmental Management Research, Inc., Texas A&M University Texas Institute of Oceanography.

2

Lightering Primer

The solid safety record for lightering in the United States, which was documented in the previous chapter, does not suggest obvious avenues for risk reduction. Improving safety at the margins requires laying a foundation through a careful review of current lightering activity patterns, government regulations, industry standards, and operational procedures and equipment. This chapter summarizes the status and technical state of the art for lightering in the United States. The first section is an overview of U.S. lightering activity, including oil import patterns, types of vessels, and the nature of activities on each coast. The second section outlines the USCG regulatory regime for lightering and industry progress in setting standards. The third section describes the current lightering process in the United States.

OVERVIEW OF U.S. LIGHTERING ACTIVITY

Lightering takes place in the Gulf of Mexico and along the east and west coasts (see Figure 1-1). Approximately 60 percent of all U.S. oil imports are delivered through the Gulf of Mexico, and about 30 percent of the total comes through the LOOP and lightering areas (NRC, 1997). Most of this crude oil is imported from the Arabian Gulf, West Africa, the North Sea, and the Caribbean Basin (see Table 2-1). In addition, some refined petroleum products are imported from the Caribbean, Canada, and Europe and lightered (along with refined products from the U.S. Gulf of Mexico) on the East Coast. The committee could not locate any data on this trade, but the volume of products lightered is very small compared to the volume of imported crude oil lightered.

The numbers tell only part of the story. The characteristics of lightering activity, as well as public perceptions of it, vary by coast and by state. In Texas,

TABLE 2-1 Sources of U.S. Crude Oil Imports Delivered by Offshore Lightering, 1994

Region	Barrels (millions)	Percentage
Arabian Gulf	351.5	59
Africa	112.8	19
Europe	64.3	11
Americas	61.2	10
Pacific	6.2	1
Total	596.0	100

Source: Maritime Administration Office of Statistical Analysis

for example, state officials recognize lightering as an integral part of the economy and invite industry to provide data and help develop lightering rules and regulations. Companies operating in the Gulf of Mexico have joined together through ITOL to develop lightering standards. On the East Coast, inshore lightering has been a common practice in Delaware Bay for many years, but offshore activities have developed only recently and tend to be more varied.

Lightering patterns in the United States are currently fairly stable (Figure 2-1). Oil imports to the United States have been rising steadily since the early 1980s and now total about 7.5 million barrels per day. The U.S. Department of Energy has projected an increase in U.S. oil consumption of 3.5 million barrels per day between 1994 and 2015 (Energy Information Administration, 1996). With more than half of U.S. demand now being met by foreign sources and most imported oil being delivered by water, the nation faces the prospect of a substantial increase in the waterborne oil trade (NRC, 1997). Lightering will undoubtedly continue to be a popular method of delivering imported oil, but predictions of future activity levels are based on considerable speculation.

Most lightering of imported oil takes place in either the Gulf of Mexico or Delaware Bay. Future imports to the Gulf of Mexico will most likely be lightered if they are carried by large tankers from distant sources, such as the Arabian Gulf or Africa. Because the proportion of long-haul versus short-haul imports is unlikely to change dramatically in the near future, the committee expects that lightering activity will remain fairly steady in the Gulf of Mexico in the near term. On the East Coast, lightering of refined products has increased in the recent past but still makes up a very small percentage of the total volume of lightered cargo. (This trade depends on the fluctuating margins between the cost of imports and the domestic supply.) East Coast oil refineries are now operating at nearly full capacity, and future increases are likely to be limited to improvements in efficiency because laws to protect air, water, and coastal zones will make the construction of new refineries cost prohibitive. Accordingly, the committee anticipates only small increases in lightering of crude oil on the East Coast in the near term. Trends on the West Coast are expected to have little impact on overall

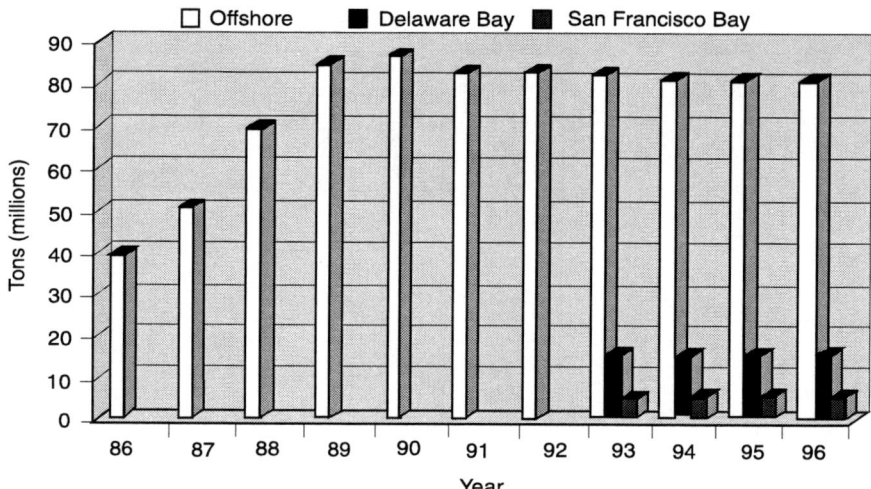

FIGURE 2-1 Crude oil deliveries to U.S. ports by lightering, 1986 to 1996.
Notes:

Offshore lightering of imported crude oil for the years 1986 to 1994 are from MARAD
Office of Statistical Analysis. More than 95 percent of these offshore imports are
lightered in the Gulf of Mexico.

Because of data analysis problems in MARAD for the years 1995 and 1996, the numbers for those years are based on rough estimates by ITOL of offshore lightering activity in the Gulf of Mexico.

Delaware Bay figures are from Maritrans Corporation's estimates for 1993 to 1996
(previous years are not available).

San Francisco Bay figures (Alaskan crude oil) were derived from local USCG estimates for 1993 to1996 (previous years were not available).

lightering activity levels, given the very small volume of oil lightered there (see
Table 1-2).[1,2]

The present study assumes that U.S. lightering activity will remain about the
same as it is now. Nevertheless, preparing for the possibility that lightering activity will increase and that the industry will expand is only prudent.

---

[1]A substantial proportion of Alaskan oil is lightered in San Francisco Bay, but domestic production
is likely to fall in the future. Oil imports may rise to compensate, but new lightering activity involving
imported oil would take place offshore.

[2]The committee reviewed several projections of import patterns and concluded that none was con-
clusive enough about their effect on lightering to establish a consistent trend. Although imported oil
is usually predicted to increase, whether that oil is lightered depends greatly on where it comes from
(short hauls from South America would probably not be lightered, and oil from Mexico would be
pipelined; oil from the Arabian Gulf or Africa would probably be lightered). In addition, the limits of
U.S. refinery capacity will restrict the amounts of imported crude oil and possibly cause a shift to
imported products, which would arrive by smaller vessels and would not be lightered.

Vessels Involved in Lightering

Ships to be Lightered

Most of the crude oil imported into the Gulf of Mexico originates on continents thousands of miles away and is usually transported in very large crude carriers (VLCCs) of 165,000 to 360,000 deadweight tons (DWT)[3] or ultralarge crude carriers (ULCCs) with capacities of up to 550,000 DWT. These tankers can be as large as 1,500 feet long and 225 feet wide, with a cargo capacity of 4 million barrels (168 million gallons). Ships of this size have deep drafts, extending as much as 90 feet below the water surface. Therefore, VLCCs and ULCCs cannot enter most U.S. ports, which typically have harbors and channels ranging from 30 to 40 feet deep. (The only U.S. port that these vessels can use is the offshore LOOP terminal. Another factor preventing their entry into U.S. ports is the sheer physical size and maneuvering characteristics of these large ships, which require ample time, broad channel widths, and turning basins.

The STBLs that supply the East Coast refineries are typically Suezmax vessels of 120,000 to 165,000 DWT (approximately 1 million barrels) or, in some cases, Aframax vessels of 80,000 to 105,000 DWT (approximately 650,000 barrels). When these vessels are fully loaded, their drafts exceed the maximum channel depths in the Delaware River (40 feet) or New York Harbor (40 feet or less, depending on the location in the port). Most of the crude oil arriving at these refining centers originates in West Africa. The remainder comes from the North Sea, the Mediterranean, or the Middle East.

Product carriers calling on East Coast ports are typically 50,000 DWT or less. Their voyages originate in the Gulf of Mexico, the Caribbean, Canada, or Europe. These vessels are lightered because of either port-specific draft restrictions or the economic advantages of lightering as opposed to having the vessel call at more than one port.

Service Vessels

Service vessels can be of several types. Vessels dedicated to the lightering trade spend the majority of their time shuttling between STBLs and terminals. Nothing restricts these vessels from participating, when lightering business is slow, in the more typical dock-to-dock transshipment of petroleum. Some dedicated vessels are owned by, or on long-term charter to, oil companies that attempt to provide a minimum level of secured lightering tonnage. Other dedicated vessels are available on a first-come, first-served basis (the "spot market") to meet

---

[3]Deadweight ton is a measure of a vessel's carrying capacity (the weight of cargo, fuel, fresh water, and stores).

the needs of oil companies with excess chartered tonnage or oil companies or oil traders who prefer to deal on a trip-by-trip charter basis. Dedicated service vessels typically are permanently outfitted for lightering and have all the necessary equipment on board (e.g., fenders, mooring lines, deck equipment, and hoses), as well as the necessary personnel. Service vessels used in the Gulf of Mexico and on the West Coast are small enough to enter port areas, usually from 80,000 to 150,000 DWT (with a cargo capacity of 375,000 to 1.1 million barrels).

The majority of dedicated tug-barge units used in lightering operate along the East Coast. The capacity of dedicated tug-barge units ranges from 50,000 barrels to 400,000 barrels. Typically, units with capacities of more than 130,000 barrels are built as integrated tug-barge units, in which the tug and barge are attached with semi-rigid mechanical devices or a combination of a deep notch and soft-line or wire-tensioning system. Smaller units are designed so that the tug is connected to the barge by a series of soft lines or wires on the stern or alongside. These connections are very sensitive to sea conditions. If the seas get too rough, the tug must tow the barge. Typically smaller barges are used to lighter refined products, whereas larger vessels are used to lighter crude oil.

Vessels that are not permanently equipped with lightering equipment are sometimes known as all-purpose vessels. All-purpose vessels may be on long-term contract to provide lightering services, or they may be chartered for one or more voyages. They can be the same sizes and types as the vessels described above; but they must be supplied with the necessary equipment prior to a lightering operation. Fenders and hoses are usually brought to an offshore lightering location by a workboat, or the equipment is placed on board while the vessel is at a dock just prior to the voyage. When these vessels are not involved in lightering, they spend the majority of their time transporting petroleum from dock to dock.

The crews and officers aboard dedicated vessels are usually highly knowledgeable about lightering in general and their vessel's capabilities in particular. The officers and crews aboard all-purpose service vessels, which are used only for an occasional lift from an STBL, may not have in-depth knowledge of lightering operations. These vessels can still conduct lightering operations safely, as long as everyone involved is aware of the circumstances and procedures and plans accordingly. Lightering companies usually supplement the crews by placing an experienced lightering master aboard the service vessel and, frequently, an assistant lightering master aboard the STBL.

Workboats / Support Vessels

Many companies conducting offshore lightering operations enlist the assistance of a workboat, especially when using nondedicated service vessels. Workboats are typically 145 to 175 feet long and are used to transport lightering equipment. If service vessels are not built specifically for lightering, workboats bring out the necessary fenders and hoses. Some workboats used in lightering are also equipped with firefighting and oil pollution response equipment. Workboats that remain on station near a lightering operation can provide first response in the event of a spill while additional equipment is being mobilized and brought to the scene. Workboats can also be used as command platforms. Although workboats are not required by regulation or law, many shipping companies consider them safety nets for offshore lightering operations. The need for workboats is highly dependent on local conditions, specific characteristics of all vessels, and equipment and operational details. The committee considers that the Coast Guard captain of the port (COTP) oversight authority will assure prudent use of appropriate vessels.

Lightering in the Gulf of Mexico

Lightering activity in the U.S. Gulf of Mexico takes place in either designated lightering zones or traditional lightering areas (see Figure 2-2). STBLs can be at anchor, under way, or drifting, depending on weather and sea conditions. Service vessels used in the Gulf of Mexico are generally smaller tankers, in the 80,000 to 100,000 DWT range. Typically, STBLs lighter their entire cargoes because most of them are too large to enter port areas even after much of the cargo has been removed and their vessel drafts have been substantially reduced. Some of the smaller STBLs carrying crude oil from West Africa can enter port after one lift by a service vessel. Three lightering service companies[4] conduct most of the operations in the Gulf of Mexico.

Nine areas have traditionally been used for lightering. Table 2-2 indicates the center point of each area. These traditional lightering areas, which can be many miles wide and long, are located away from the busy fairways or traffic separation schemes leading into port areas and away from large concentrations of offshore exploration and production platforms. Mariners who often navigate these waters have come to expect lightering operations in the traditional areas and can be expected to take appropriate precautions when approaching or transiting known lightering areas.

The USCG was authorized by OPA 90 to create designated lightering zones, where single-hull vessels are permitted to operate for a period of time. Until January 1, 2015, a tank vessel need not comply with the double-hull requirement if it is either "(A) a vessel unloading in bulk at a deepwater port licensed under the Deepwater Port Act of 1974; or, (B) a delivering vessel that is offloading in lightering activities within a designated lightering zone established under title 46 USC section 3715(b)(5) and more than 60 miles from the baseline from which the territorial sea of the United States is measured." By using the designated

---

[4]The three companies are Skaugen PetroTrans, Inc., OMI Petrolink Corporation, and American Eagle Tankers, Inc.

34

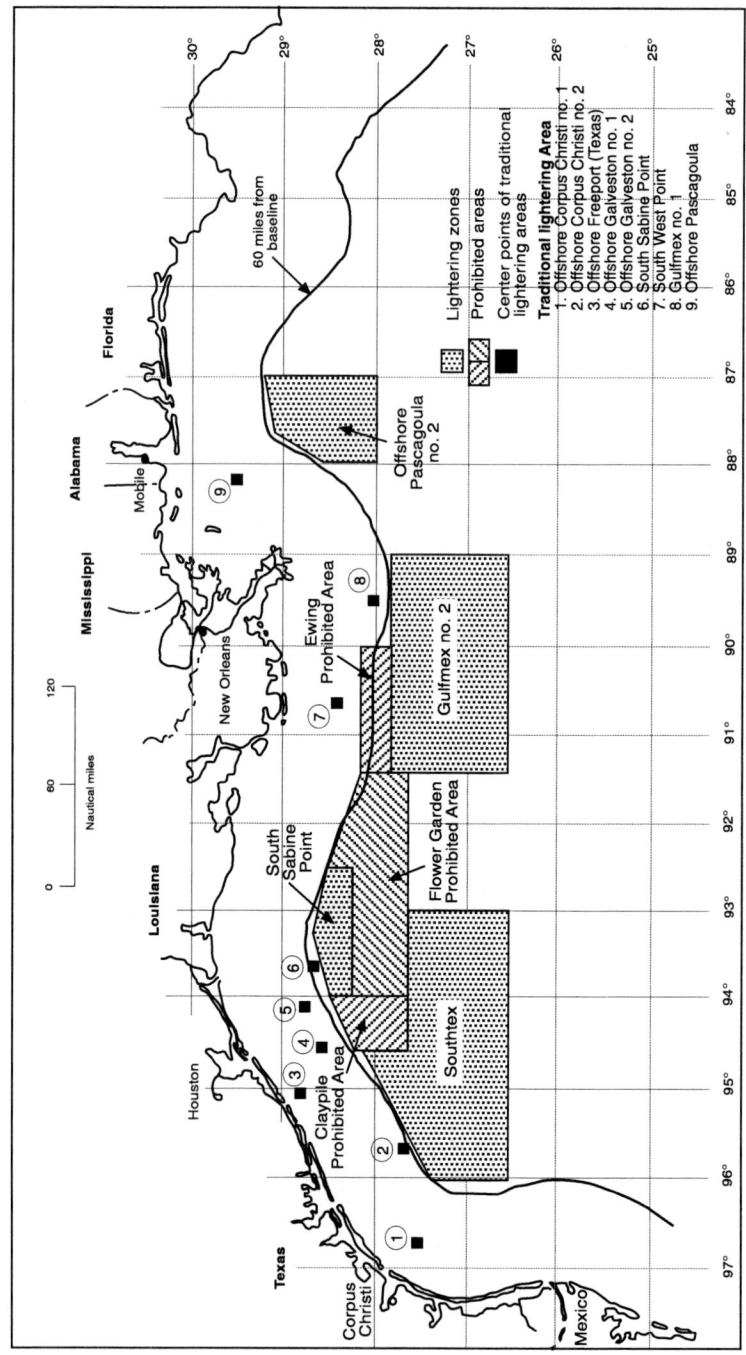

FIGURE 2-2 Designated lightering zones and prohibited areas in the Gulf of Mexico.

TABLE 2-2 Traditional Lightering Areas in the U.S. Gulf of Mexico

Area	Latitude	Longitude
Sabine, Texas	28°–30°N	93°–40°W
Galveston, Texas	28°–27°N	94°–30°W
Galveston (#2), Texas	28°–40°N	94°–08°W
Freeport, Texas	28°–45°N	95°–03°W
Corpus Christi, Texas	27°–28°N	96°–49°W
Corpus Christi (#2), Texas	27°–48°N	95°–31°W
Houma, Louisiana	28°–27°N	90°–42°W
Pascagoula, Mississippi	29°–27°N	88°–13°W
Southwest Pass, Louisiana	28°–00°N	89°–30°W

lightering zones, single-hull tank vessels contracted for after June 30, 1990, and older single-hull tank vessels phased out under OPA 90, will be able to lighter until 2015.[5] At the time these rules were made, approximately 20 new vessels meeting this definition were identified by the USCG that might trade to the U.S. Gulf of Mexico and engage in lightering to offload cargo. Because the International Convention for the Prevention of Pollution from Ships (MARPOL Convention) prohibits, as of July 1993, contracting for the construction of single-hull tankers used in international commerce, no more single-hull tankers will be built. The intent of allowing single-hull tankers to continue to lighter in designated areas was to ensure adequate oil supplies in the United States while double-hull tank vessels were being built and placed in service.

The four lightering zones shown in Figure 2-2 were formally designated by federal regulation (46 USC 1503[b][5]) on August 29, 1995. These are the only lightering zones currently designated by national regulation, although other regulated areas can be (and are) designated at the USCG district level.[6]

Lightering on the East Coast

Most lightering along the East Coast takes place in protected and semi-protected coastal areas, primarily Big Stone Anchorage in Delaware Bay and, to

---

[5]Single-hull tank vessels contracted for after June 30, 1990, are prohibited from conducting lightering operations anywhere (other than the designated zones) in the U.S. Gulf of Mexico, unless they are more than 200 miles offshore, outside the U.S. Exclusive Economic Zone (EEZ). The prohibition on operations inside the EEZ raises concerns that lightering operations might simply move farther offshore into international waters where the U.S. has no jurisdiction.

[6]The Coast Guard COTP can use his general authority to designate anchorages or regulated navigational areas with respect to lightering within U.S. waters. For example, this has been done in Long Island Sound, Delaware Bay, and San Francisco Bay. When a vessel provides the required advance notice of lightering to the COTP, he can require that the vessel use a specified area before giving permission.

some extent, in Long Island Sound, New York Harbor, and Boston Harbor. Lightering activity is concentrated inshore (see Table 1-2),[7] which influences the types of vessels used. Under federal law, petroleum imports that are lightered within U.S. territorial waters must be transferred to U.S.-flag service vessels (OTA, 1989), which are subject to different standards and regulations than vessels engaged in offshore lightering (see section on government and industry controls below).

In these inshore areas, the STBL is always at anchor, and barges or integrated tug-barge units are usually used to transport cargo to the discharge berth. Vessels lightering in semiprotected areas can be subjected to weather anomalies, such as high seas, swells, currents, and winds. The weather is less of an issue in protected bays, harbors, and sounds. The STBLs used in crude oil lightering are generally smaller than those used in the Gulf of Mexico, typically carrying about 1 million barrels of oil. In some instances, after a few lifts by service vessels, the draft of the STBL is sufficiently reduced for it to move to the dock and directly offload the remaining cargo.

Refined products (i.e., gasoline, jet fuel, and diesel fuel) are lightered in Long Island Sound and various East Coast harbors. The STBLs are typically 30,000 to 50,000 DWT tankers originating at refineries in the Caribbean, the Gulf of Mexico, Canada, or Europe. In these cases, lightering is performed either because the vessel's draft is too deep to enter the destination port directly or because the product is destined for two or more terminals, and it is more economical to lighter the parcels than to make multiple port calls.

Offshore lightering is conducted off Montauk Point, New York; Cape Henlopen, Delaware; Cape Henry, Virginia; and Great Issacs, Bahamas. Lightering is conducted off Great Issacs when an STBL carries cargo that needs to be delivered at both East Coast and Gulf Coast refineries or when the weather off Cape Henry is not conducive to lightering operations. Offshore lightering on the East Coast is subject to different weather and wave conditions than those found along the other U.S. coasts. Along the East Coast, storms and rough seas are more prevalent in the winter, spring, and fall seasons than in the Gulf of Mexico, but the large swells that occur offshore in the Pacific (off the West Coast) are not as prevalent along the Atlantic coast.

In most East Coast locations where lightering takes place, the local USCG COTP has substantial control over lightering operations. A COTP has the authority to establish certain rules and regulations within the territorial sea under the Ports and Waterways Safety Act (33 USC section 1221 et seq.). Notification is required prior to a lightering operation (33 CFR 156.215). A COTP can also designate a lightering anchorage (46 USC section 3715), as COTPs have done in Stapleton in New York Harbor, and President Roads in Rhode Island. Another

---

[7]Maritrans handles most of the lightering on the East Coast; Skaugen is beginning to provide offshore lightering services here as well.

approach is to establish a regulated navigation area (33 CFR 165) within the territorial sea, such as Big Stone Anchorage in Delaware Bay. Recently, the COTP in Long Island Sound proposed that the USCG district commander establish five regulated navigation areas in waters that are currently used for lightering. The areas were recommended initially by a stakeholder group composed of industry representatives and other waterway users. The areas were reviewed by the USCG based on environmental protection and traffic safety criteria. The regulated navigation areas would be designated through a rule-making process to formalize the applicable federal regulations (personal communication from Capt. Peter Mitchell, December 15, 1997). An overview of government and industry controls on lightering is provided later in this chapter.

Lightering on the West Coast

Substantial lightering takes place in only two locations on the West Coast: in San Francisco Bay and about 140 miles off San Diego. Small amounts of petroleum are also lightered in Puget Sound and in Los Angeles/Long Beach harbor, but not on a regular basis. In the past, some lightering was done offshore near Los Angeles (inside the Channel Islands), but this practice has been discontinued.

In San Francisco Bay, most of the lightering involves shipments of crude oil from Valdez, Alaska, to the refineries in the San Francisco Bay area.[8] The protected area offers some shelter from weather conditions. Typically, tankers of about 150,000 DWT anchor in a designated area (Anchorage #9) just south of the Bay Bridge and discharge cargo into barges and other vessels, which then navigate a shallow channel to the north (with draft restrictions of about 35 feet) to deliver the cargo to refineries. Lightering is one of the most cost-effective ways of bringing Alaskan oil to these refineries. Only one major refinery in the region is located along channels deep enough[9] to allow large Alaskan trade tankers to dock.

The lightering activity off San Diego began in 1996. Chevron Shipping Company, the only company now operating in this region, uses lightering to bring crude oil from the Middle East and the Far East to refineries in the Los Angeles and San Francisco areas.[10] The typical STBL is a VLCC of about 300,000 DWT that lighters into service vessels of about 150,000 DWT, which then deliver crude oil to Chevron refineries. These operations take place in unprotected waters where weather conditions can be more severe than in most other U.S. locations where

---

[8]One company, SeaRiver Maritime, conducts more than 80 percent of the lightering in San Francisco Bay; the rest is performed by approximately six smaller companies.

[9]These channels are still only 39 feet deep, and most vessels have to either wait for high tide or offload some cargo before going in.

[10]The state of California has commented on the Chevron operations (see Appendix F) even though they take place beyond state jurisdiction. The Coast Guard monitors this offshore lightering, as it does in the Gulf of Mexico.

lightering now takes place. Accordingly, Chevron is developing special (more rugged) equipment and procedures to accommodate severe weather. The company uses dedicated service vessels and has tested and trained crews specifically for these operations.

California officials and local environmental groups monitor West Coast lightering operations, and, impressed by the good safety record to date and by industry's cooperation in information sharing, they did not express serious concerns about the risk of spills from lightering. In the future, lightering activity levels are expected to remain stable unless and until there are major changes in the Alaskan oil supply. Some have predicted a decline in Alaskan crude oil production in the next several years; if this occurs, then more foreign crude oil will be imported to supply West Coast refineries. An increase in imports could lead to more lightering because foreign oil would most likely be delivered in VLCCs and would have to be lightered to smaller service vessels offshore. Other changes in crude supply patterns, such as an increase in deliveries from South America, could have similar effects.

GOVERNMENT AND INDUSTRY CONTROLS

The only federal agency that oversees lightering operations beyond the territorial seas is the USCG, whose role varies depending on the geographical location of the lightering operation. The USCG exerts maximum control over lightering within the territorial sea, where all service vessels must be U.S.-flag vessels, constructed in a U.S. shipyard, and crewed by U.S. citizens. Lightering outside the territorial sea can be performed by foreign-flag vessels built overseas that employ foreign crews.

In either case, the USCG exerts varying control over lightering through six general mechanisms: vessel design requirements (established by the *United States Code*, OPA 90, and international agreement)[11]; operational procedures (established by the *United States Code*, OPA 90[12] and international agreement)[13]; personnel qualifications; oil spill contingency planning and equipment requirements; vessel inspection; and monitoring. The USCG role in these areas is outlined

---

[11]This agreement is the MARPOL Convention, which is the 1978 Protocol of the International Convention for the Prevention of Pollution from Ships (1973). The MARPOL Convention is one of many conventions adopted by the International Maritime Organization (IMO), the United Nations agency responsible for maritime safety and protection of the marine environment. All of the world's major shipping nations are members of IMO.

[12]For example, one regulation promulgated under OPA 90 requires vessel masters of single-hull tankers to calculate the underkeel clearance based on the vessel's draft and other conditions prior to entering port, but there is no established minimum clearance (33 CFR Part 157.455).

[13]This agreement is the International Convention on Standards of Training, Certification, and Watchkeeping for Seafarers (known as STCW), which is discussed further in Chapter 4.

briefly below. The relevant vessel design issues, equipment requirements, and personnel qualifications are addressed in more detail in chapters 3 and 4.

There are two sets of basic USCG regulations; one for navigable waters and contiguous zones (i.e., inshore areas) and one for offshore areas. The specialized USCG lightering zone regulations (33 CFR Part 156 Subpart C) apply only to the four designated offshore lightering zones in the Gulf of Mexico (the regulations are provided as Appendix G). Elsewhere, lightering is performed at the discretion of the local COTP within the parameters of the applicable basic lightering regulations and other applicable laws. Table 2-3 outlines the various regimes for USCG control of lightering operations.

All STBLs conducting operations within the U.S. EEZ, a 200-mile-wide band around the coastline, must notify the nearest COTP more than 24 hours prior to the vessel's arrival. This applies to operations in both designated lightering zones and traditional lightering areas. The vessel notifies the USCG of both intent to lighter and the proposed location, and the COTP grants permission subject to certain guidelines and standards. Various areas (e.g., Big Stone Anchorage in the Delaware River and Graves End Bay and Stapleton Anchorage in New York Harbor) have been designated as lightering anchorages in which standard operating procedures must be followed. The regulations may be suspended when lightering is carried out during emergency salvage operations.

For vessels lightering in a U.S. coastal region within 200 miles, a number of international and national regulations apply. For example, both the STBL and service vessel must have a U.S. certificate of financial responsibility (COFR) and the minimum level of liability insurance established by OPA 90. The vessel owner, not the cargo owner, is liable for any spills. Both vessels must also have on board USCG-approved spill-response plans that identify procedures, equipment, and personnel available to respond to an oil spill (see Box 2-1). All U.S.-

TABLE 2-3 Regimes for U.S. Coast Guard Control of Lightering Operations

Geographical Area	Designator	Location	Type of Control
Designated lightering zones	U.S. Coast Guard	Gulf of Mexico	Special federal regulations
Traditional lightering areas	Local shipping industry	Various offshore areas on all coasts	Federal regulations for offshore activity
Lightering anchorages	Captain of the port (U.S. Coast Guard District)	Delaware Bay, New York Harbor, San Francisco Bay, et al.	Federal regulations for inshore transfers
Regulated navigation areas	Captain of the port (U.S. Coast Guard District)	Long Island Sound, Delaware Bay	Established by captain of the port
Outside U.S. Exclusive Economic Zone (EEZ)	U.S. Congress designates the EEZ	More than 200 miles off all U.S. coasts	No U.S. Coast Guard control

---

BOX 2-1 Requirements for Spill Response Plans

The OPA 90 required the development of regulations for response capabilities for oil spills. The USCG established regulations (33 CFR Part 155 Subpart D–Response Plans) accordingly. As of 1995, all vessels engaging in lightering operations within the EEZ, when the cargo lightered is destined for a port or place subject to the jurisdiction of the United States, must have a USCG-approved response plan. The plans must identify the personnel and equipment, secured by contract or other approved means, that would be used in the event of an oil spill, as well as certain shipboard requirements for personnel training, equipment, and drills.

The regulations also specify response times and amounts of equipment and oil-recovery capacities, depending on the location, size, and type of oil spilled. For example, the lowest planning standard (for the "average most probable" spill) is defined as "a discharge of 50 barrels of oil from the vessel during a transfer operation." All vessels transferring oil within 12 miles of land must have the capability to deploy a "containment boom in a quantity equal to twice the length of the largest vessel involved in the transfer and capable of being deployed at the site of the oil transfer operation within one hour of detection of a spill; and oil recovery devices and recovered-oil storage capacity capable of being at the transfer area within two hours of the detection of a spill during the transfer operation."

A waiver granted by the USCG commandant establishes an alternative planning standard for vessels engaged in lightering operations more than 12 miles from shore. This planning requirement allows one hour for mobilization of equipment and a planned transit speed of 5 knots to the site of the transfer operation.

---

flag vessels must have a certificate of inspection. Foreign tank vessels must have a valid tank vessel examination letter (TVEL) issued by the Coast Guard. All vessels must comply with all USCG regulations covering the operation of tank vessels.

Vessels engaged in lightering in the territorial sea are required to abide by all regulations pertaining to the transfer of oil. These regulations address several parameters, such as who is in charge, discharge cleanup, the connection of transfer hoses, and equipment tests and inspections. Prior to a transfer, both vessels must complete a Declaration of Inspection (DOI) (46 CFR 35.35-30) and are subject to all of the applicable provisions (in 46 CFR Subchapter D–Tank Vessels and 33 CFR Subchapter O–Pollution). The DOI requires that the individuals in charge of a petroleum transfer attest to their compliance with a predetermined set of regulations. A DOI is also required for offshore lightering (33 CFR Part 156.210).

Most vessels involved in offshore lightering are foreign-flag vessels (OTA, 1989) because they are less expensive and more readily available than U.S.-flag vessels and, under the law, can be used for lightering offshore. Between the territorial sea and the 200-mile limit of the EEZ, the USCG's primary tool for ensuring safety is the requirement that foreign-flag vessels engaged in lightering have a current TVEL certifying that the USCG has inspected the vessel to international standards and certain U.S. tanker requirements within the last 12 months. If the TVEL is not current, then the USCG must conduct an inspection before lightering begins.

The designated lightering zones in the Gulf of Mexico provide the USCG with some additional enforcement tools in those areas. For example, the regulations describe the weather conditions in which lightering can be performed and specify the maximum operating conditions, work hours, and operational parameters for vessels engaged in lightering in the designated areas. Although these regulations do not apply to traditional lightering areas, most vessels operating in those areas voluntarily follow the regulations for designated zones. Furthermore, most ITOL guidelines for operational parameters are stricter than the regulations.

As a further safety check in the Gulf of Mexico, the USCG engages in periodic surveillance of lightering areas. Approximately three to four times a week, air crews stationed in Corpus Christi, Texas, fly over lightering areas. Sometimes they have a specific mission, but usually they just observe whatever vessels happen to be in the lightering areas. Anecdotal reports suggest that USCG personnel sometimes observe lightering operations on other coasts.

The shipping industry has implemented lightering safety measures in several stages. In the early 1970s, large fenders and special mooring arrangements were evaluated and adopted (NRC, 1991). The procedures for maneuvering vessels, mooring, and handling cargo were developed by the OCIMF (Oil Companies International Marine Forum), an oil industry association, and codified in *Ship to Ship Transfer Guide*, first published in 1978 and now in its third edition (ICS and OCIMF, 1997). These guidelines are for offshore lightering between tankers only. There are no specific lightering guidelines or standards specifically for inshore lightering, which often involves barges, although the American Waterways Operators promotes general safety standards and practices for inland vessels (see Chapter 4). Many inshore operators follow the international guidelines. The committee considers the international guidelines to be sufficiently stringent for general application, and the guidelines also encourage specific enhancements under special conditions that warrant them.

In the late 1980s, the COTP in Houston became concerned about lightering activities in the Gulf of Mexico, which led to the establishment of ITOL in 1990. ITOL is a mechanism for the consensus-based development of industry standards, and among its accomplishments, is the publication of a supplement to the international guidelines on lightering operations (ITOL, 1990). In addition, some major oil companies and independent oil transport companies have established their own lightering procedures, which are generally more stringent than the international guidelines.

THE LIGHTERING PROCESS

The crude oil lightering process is initially set in motion by a refinery's capabilities to refine certain types of crude oils, which are distinguished by their source and characteristics. The refinery's capabilities determine where the oil will be shipped. For example, a refinery designed to run a light, sweet crude oil[14] would probably get it from the Arabian Gulf rather than from Mexico. The supply planners attempt to make 30-to-60-day projections of the product lines to be run, based on the demand, season, and refinery capability. The projections dictate not only the source of the oil but also the amount needed. Product availability and economics also affect the final determination of the source and cost of the oil, as well as the transport costs and the spot deals possible on the open market. Existing contracts and agreements for delivering a certain number of barrels per month may also affect the source of the oil.

Once the refinery plans are established, the traders working under established agreements or on the open market purchase the necessary "raw material barrels." The traders begin with the projected dates on which the oil will be needed at the refinery and calculate back to an estimated transportation date. Then, based on the traders' purchases, the vessels needed to transport the oil are secured by the traders. These vessels are either managed by the oil company or are chartered for a specific voyage or for a predetermined length of time (independent of the number of trips). The types of companies that provide lightering services are described in Box 2-2.

Refined products are typically imported by oil companies or traders either when the domestic supply can not meet the demand or when the margin is better on the imported product even after shipment costs are added. Lightering of petroleum products almost always takes place inshore, typically using a tug-barge unit as the service vessel. The STBL can be either a tanker or a large tank barge.

The lightering process can be divided into the approach, transfer, and post-transfer (or departure) phases. The approach phase includes the rendezvous in the lightering area or zone, the maneuvering of the STBL (if not already at anchor) and service vessel alongside each other, the mooring of the two vessels, and (if necessary or desirable) the anchoring of the STBL. The transfer phase includes the completion of required paperwork, the connection of hoses, the pumping and transfer of cargo, and the inspection and certification of cargo. The post-transfer phase includes operations associated with the unmooring and departure of the service vessel. In certain areas (e.g., the designated lightering zones), the approach distances, wave heights, weather conditions, and other factors are

---

[14]A light, sweet crude oil is a low-density, low-viscosity oil with a low sulfur content.

BOX 2-2 Companies That Provide Lightering Services

Service companies can provide any or all lightering equipment and personnel. When hired to support lightering operations, service companies usually provide the full range of services because this is the most economical arrangement for the customer.

Some of the larger oil companies retain one or more service vessels on permanent charter to service STBLs on a regular basis, supplemented by additional chartered service vessels as necessary. The oil companies usually scrutinize every aspect of a lightering operation very carefully and, if incidents occur, they act quickly to ensure that they do not happen again. They may stop doing business with a poor performer.

One major oil company owns and uses two dedicated tankers to service STBLs carrying oil to its refinery. The company uses its own staff mooring and lightering masters when necessary.

---

spelled out, but in general there are no specific federal regulations governing the approach, mooring, unmooring, and coming off procedures (other than specified approach distances and transition positions). The best practices for offshore lightering have been established by the OCIMF (ICS and OCIMF, 1997). An annotated timeline for the overall oil transportation and lightering process is provided in Chapter 4 (Box 4-1 and Figure 4-1).

Approach Phase

Offshore Operations

The approach phase begins when the service vessel proceeds to the general area where the STBL is waiting. Communications are established and maintained throughout the operation among the vessel masters, the mooring (or lightering) master, and the assistant mooring master, if present. (The personnel involved in lightering operations are described in Box 2-3.) By tradition and custom, everyone involved in the approach phase must agree that the conditions, vessels, equipment, and personnel are prepared and adequate for a safe operation; otherwise, the process comes to a halt until the problems are resolved. Several OCIMF safety checklists (copies are provided in Appendix H) are then completed before any close-quarters maneuvering is done to ensure that both vessels comply with international guidelines and company policies. Prior to inshore lightering, a DOI must be filled out and transmitted to the COTP. The committee considers that the safety checklists under the 1997 guidelines and the current lightering zone regulations noted in Box 1-3 have adequately addressed the problems of "limited communications."

BOX 2-3 Personnel Involved in Lightering Operations

Vessel masters: The masters of the STBL and service vessel retain ultimate authority over their vessels during the entire operation, even when a mooring and lightering master are on board to assist with the operation. Depending on a vessel's typical route, the master may or may not be familiar with lightering procedures. A service vessel owned by a company that provides lightering services may have a master who is very familiar with lightering. Masters of STBLs that are regularly chartered for lightering operations usually are somewhat familiar with the operation.

Mooring master or lightering master: A mooring master (or lightering master) is an experienced, senior officer placed on board the service vessel to coordinate and assist in directing the entire lightering operation, from the first communications during the approach phase to the coming-off process. The mooring master is especially valuable for monitoring weather conditions, assisting with navigation, and ensuring a safe operation.

Assistant mooring master: Sometimes an assistant mooring master is placed a board the STBL to ensure that the mooring master's directions are followed and to assist the crew. This individual could be a mooring master in training.

Cargo mate, chief mate, or chief officer: One or more persons are on board both the STBL and service vessel to ensure the safe loading and discharge of cargo within the specified parameters of stress, trim, and stability.

Workboat master: A workboat is responsible for transferring fenders and hoses and usually remains in the lightering area during the entire operation. A workboat often heads off small boat traffic that might interfere with the operation and may also provide support for responding to oil spills, if necessary.

Officers and crews of STBL, service vessel, and workboat: The crews perform various duties, such as handling the lines that moor the two vessels together, standing watch on deck to ensure that cargo is transferred smoothly, providing navigation watches, and ensuring that the engines are ready for maneuvering.

During lightering operations, the service vessel is usually moored to the starboard side of the STBL. Before the vessels begin to maneuver to come alongside, a string of four fenders is rigged on either the starboard side of the STBL or on the service vessel (the industry-recommended approach). The rubber fenders, ranging in size from 3 by 6 feet to 15 by 30 feet, are filled with either air or foam. Primary fenders, typically of the larger size, float on the water surface to cushion the impact as the two vessels are brought together. These fenders must remain afloat throughout the entire operation to prevent their riding up the sides of the vessels and rolling onto the deck. Smaller secondary fenders (also called baby fenders) are used, as necessary, on either the STBL or service vessel to prevent unexpected bow or stern contact during mooring or unmooring. The fenders are brought out to the STBL by either the service vessel or a workboat.

When the vessels are approximately 25 to 30 feet apart, the crews begin passing mooring lines between them to make them fast to one another. As the two vessels come closer together, the mooring master must pay special attention to the hydrodynamic effects that tend to move the bow of the service vessel away from the STBL; therefore, the first lines out are the forward spring lines. The head lines follow shortly thereafter. Once these critical lines are out, a full complement of after-spring lines and stern lines is put out.

When the lines are almost fast, the mooring master begins to reduce the speed of both vessels simultaneously. If the lightering operation is to take place at anchor, then the service vessel's speed is gradually reduced to zero, essentially making it a towed vessel. The STBL proceeds slowly to the chosen anchorage, and the anchor is dropped. If the two vessels are to drift with the winds and currents, then the speed of both is reduced to zero. The third option is to lighter while the STBL steams at a very slow speed and the service vessel is towed alongside. The selection of one of these three alternatives depends on several variables, such as water depth, weather conditions, sea states, the area available for steaming, and the draft of the STBL. In all cases, the mooring master, or lightering master, and masters and crews of both vessels must maintain diligent navigational watches as long as the vessels are together to ensure the safety of the crews and vessels. Support vessels, such as workboats, may remain in the area to assist in the lightering operation or warn other vessels in the vicinity.

Inshore Operations Using a Tug-Barge Unit as the Service Vessel

The approach phase of an inshore lightering operation, in which a tug-barge unit is typically used as the service vessel, differs in certain respects from the initial phase of an offshore operation. First, the STBL is always at anchor when lightering takes place. Second, the interplay of a number of factors that affect how the STBL acts on its anchor must be considered prior to the approach. The approach phase begins only after the service vessel personnel have considered the wind direction; speed, direction, and time of a change in a tidal or seasonal current; vessel traffic conditions; and anchorage congestion. Third, barges are typically equipped with permanent lightering fenders, which are lowered out of fender slides and into the water prior to coming alongside the STBL. The side of the barge that berths alongside the STBL depends on how the barge is outfitted and which side provides the best shelter from the wind. Most barges are also permanently equipped with the necessary hoses and mooring lines, eliminating the need for a workboat. Fourth, mooring masters are not typically employed by the tug-barge service vessels. Rather, the tug master is responsible for directing the vessel and is assigned to the unit on the basis of proven expertise in handling vessels. Many companies that operate service vessels employ a lightering coordinator, who is stationed on the STBL. This individual's role is similar to the role of an assistant mooring master.

Transfer Phase

Offshore Operations

In preparing for the transfer of oil, pretransfer checklists and a DOI must be completed to ensure that the crews and officers of both vessels understand the amount of cargo to be transferred, transfer rates, and the expected duration of the cargo transfer. The checklists and DOI also help ensure that customary safety precautions have been taken aboard both vessels, that communications have been established between the cargo officers of both vessels, and that contingency plans have been made in case cargo is spilled.

While the checklists and DOI are being reviewed by the cargo officers, the crews hook up the transfer cargo hoses to each vessel's cargo manifold. These hoses, usually supplied by the service vessel or workboat, are constructed to withstand the pressures and transfer rates of cargo discharge. If the service vessel is equipped specifically for lightering, then the hoses are stored on board on a special mechanical arm that can span the distance between the two vessels. If the hoses are supplied by a workboat, then it comes alongside the STBL, which uses its cargo crane to lift the hoses aboard before mooring. Generally two hose strings, sets of three interconnected hoses amounting to 90 feet of hose, are used to transfer the cargo. Once the hoses have been connected and the checklists completed, cargo transfer begins.

Cargo transfer and crew communication during lightering are much the same as they are at a berth (OCIMF, 1995). The STBL starts its cargo pumps slowly as the integrity of the systems aboard both vessels is checked and rechecked. The rate of transfer is increased slowly to the maximum discharge rate. Transfer continues at this rate until the cargo tanks on the service vessel are almost full. The rate of transfer is then slowly reduced until the designated fill point is reached. The transfer is stopped when the last cargo tank on the service vessel has been topped off.

After the cargo transfer is completed, the cargo hoses are drained and disconnected from one vessel's manifold (generally the STBL). While still connected to the service vessel's manifold, the hoses are suspended vertically by a crane, so that remaining cargo runs into that vessel's tanks. (Hoses may be drained into either vessel, depending on whether a flow boom or ordinary hose rig is used, but it is preferable to drain any remaining cargo into the receiving vessel to maximize the amount offloaded.) The hoses are then capped and either draped over the side of the STBL or, if all lifts have been completed, laid on the deck in preparation for unmooring.

Inshore Operations Using a Tug-Barge Unit as the Service Vessel

An additional step is taken during the transfer phase of an inshore lightering operation. Prior to the transfer of any cargo from a vessel engaged in an international voyage to a U.S.-flag vessel engaged in a domestic voyage inside U.S. Customs waters (12 miles from shore), the STBL must undergo a customs inspection and clearance procedure. If the transfer occurs more than 12 miles offshore, and the STBL has not cleared customs, then the service vessel has completed a foreign voyage and must undergo inspection and a customs clearance process when it reenters U.S. Customs waters. As with offshore lightering a DOI must be completed prior to any petroleum transfer within the territorial sea.

Post-Transfer Phase

Offshore Operations

Cargo inspection, or gauging, is either performed while the hoses are being drained and disconnected or is delayed until the service vessel reaches its destination. Often the process is carried out at both points. At least one gauging is necessary to determine the amount of cargo transferred, ensure that each party receives the amount of cargo contracted for, and confirm that no excess cargo has been inadvertently left aboard the STBL. The cargo inspectors work on behalf of the chartering and purchasing organizations.

Some experienced mariners question the need for cargo inspections at sea, which can take more than two hours and can pose a safety risk in adverse or deteriorating weather conditions, when the mooring master is anxious to separate the vessels. Moreover, if at-sea cargo gauging takes place while the service vessel is moving in the seaway because of wind and sea conditions, the calculations of cargo levels will only be approximate. A more accurate figure can be obtained when the service vessel reaches port. One way to deal with these concerns would be to limit at-sea cargo gauging to the STBL and to begin the process only after the departure of the service vessel. Cargo levels on the service vessel can be measured at the destination port and transmitted to the STBL for record-keeping purposes. (This approach is already the norm for inshore lightering when tug-barge units are used as service vessels.) If there were concerns about possible irregularities, then a surveyor could ride the service vessel and use other methods to make initial estimates of the volume of cargo transferred.

Once the necessary cargo gauging has been performed, the mooring master or lightering master and the masters of the two vessels discuss the unmooring process, and, when all agree, the maneuver proceeds. The same level of care must be taken during unmooring and coming off as is taken when the vessels are brought together and moored. One final checklist is completed, and any necessary adjustments are made to facilitate coming off.

The mooring lines are let go in a predetermined order, and the vessels are separated very slowly. If the STBL is at anchor, then she may remain at anchor while the service vessel maneuvers. As the distance between the vessels grows, the speed and angle of departure of the service vessel can be increased. Finally, the service vessel proceeds to the discharge port and berth.

If the STBL expects another service vessel, then the crew proceeds with any cargo consolidations, crude oil washing, ballasting, or other activities required before the next lift. If no more lifts are expected, then the cargo hoses are disconnected and capped and, along with the fenders, are either collected by the service vessel or lowered to a workboat that comes alongside. The STBL then proceeds either to the next load port or to a destination port to discharge the remaining cargo.

Cargo Delivery to Shore Facilities

Following the lightering operation, the service vessel proceeds to a crude oil terminal ashore. There are 163 refineries located in the United States, many of them with either direct access or pipeline access to the sea, allowing the facility to operate with imported crude oil.

The states bordering the Gulf of Mexico account for 44 percent of all U.S. refining capacity. Galveston Bay and the Mississippi River have 11 percent of refining capacity each. Four major pipelines that supply inland refineries are located in the western Gulf of Mexico (Capline at St. James, Louisiana; Arco Pipeline at Texas City, Texas; Seaway Pipeline at Freeport, Texas; and Mobil Pipeline at Port Arthur, Texas). Other areas in which lightering takes place include Delaware Bay, with 6 percent of U.S. refining capacity, San Francisco with 5 percent, and Los Angeles with 5 percent (Oil and Gas Journal Data Book, 1998).

The port operations for a tanker mirror the lightering process in that they have an approach phase, a transfer phase, and a post-transfer phase. The vessel has to consider many factors, including port restrictions, such as draft, length, and beam. USCG regulations require the ship's master to compute underkeel clearance before entering port.

The approach phase includes maneuvering into the port area to pick up a pilot. This may well be the most dangerous time for a ship in the entire passage. Traffic, water depths, and excessive background lights are among the problems vessel masters can encounter. After picking up a pilot, the ship transits to a berth. The transit can take from 90 minutes at Freeport, Texas, to 28 hours at Baton Rouge, Louisiana.

At the terminal, the ship ties up with the assistance of tugboats. The mooring arrangement depends on the configuration of the dock. Most crude terminals use chicksans (metal arms) instead of flexible hoses for discharging the oil. Once a ship clears customs, it is gauged again, and discharge commences. Typically, ships discharge their entire cargo in less than 24 hours. During the discharge, crude oil tanks are washed to minimize retained cargo. After discharge, the cargo tanks are checked again, and pilots and tugs move the vessel back to sea.

Inshore Operations Using a Tug-Barge Unit as the Service Vessel

The post-transfer phase for inshore lightering differs in two fundamental respects from the process for offshore lightering. First, cargo gauging of the service vessel is typically done after the barge reaches the discharge dock. Second, the tug master, as always, is responsible for maneuvering the service vessel.

SUMMARY

The practice of lightering is well developed, and a safety net of industry guidelines and government controls has been established. This operational framework appears to function effectively overall, as evidenced by the rarity of spills attributed to lightering in U.S. waters.

The committee identified one gap in the framework that may offer opportunities for reducing safety risks further. The concerns expressed by some mariners about unnecessary delays caused by cargo gauging at sea appear to have merit. Moreover, the successful experience of inshore operators who have delayed cargo gauging on barges until they reached port suggests that this approach may be an effective, and perhaps safer, alternative to at-sea gauging.

REFERENCES

Energy Information Administration. 1996. Annual Energy Outlook 1996. Washington, D.C.: U.S. Department of Energy.

ITOL (Industry Taskforce on Offshore Lightering). 1990. Industry Lightering Operations Supplement to OCIMF Ship to Ship Transfer Guidelines for U.S. Gulf Coast—Galveston Zone. Houston: ITOL.

ICS (International Chamber of Shipping) and OCIMF (Oil Companies International Marine Forum). 1997. Ship to Ship Transfer Guide (Petroleum). London: Witherby & Co., Ltd.

NRC (National Research Council). 1991. Tanker Spills: Prevention by Design. Washington, D.C.: National Academy Press.

NRC. 1997. Double-Hull Tanker Legislation: An Assessment of the Oil Pollution Act of 1990. Washington, D.C.: National Academy Press.

OTA (Office of Technology Assessment). 1989. Competition in Coastal Seas: An Evaluation of Foreign Maritime Activities in the 200-Mile EEZ: Background Paper. OTA-BP-O-55. Washington, D.C.: U.S. Government Printing Office.

Oil and Gas Journal Data Book. 1998. Houston, Texas: Penn Welling Publishing.

Oil Companies International Marine Forum (OCIMF). 1995. Prevention of Oil Spillages through Cargo Pumproom Sea Valves. London: Witherby & Co., Ltd.

3

Lightering Vessels, Systems, and the External Environment

Physical factors, both on board and external, substantially affect lightering safety. Some of these factors are difficult to control. For example, industry experts say a significant variable that is difficult for lightering companies to control is the quality of the STBL. A variety of owners, operators, and vessels are engaged in transporting imported oil and thus are lightering STBLs. Cargo owners commonly contract with independent lightering service companies to provide equipment and supervisory services during lightering operations. These service companies may also operate the service vessels that shuttle cargo to refineries, or they may perform their services using chartered or independently operated vessels. The STBLs are sometimes owned and operated by the cargo owner, but more frequently the cargo owner charters the STBL. The lightering service company, therefore, often has little or no control over the quality of the STBL or the aspects of the operation that have been planned and executed by outside parties. The issue, then, is how to encourage universal adherence to the best industry standards for vessel design, equipment, operation, and maintenance.

Vessel design and the condition of the lightering equipment are critical safety factors. The inherent design of the vessel, including a single or double hull, is one factor. A strong, well designed mooring system, which keeps the STBL and service vessel together, is another. Typically, lines from both vessels are used. A fender system, suitable for each individual lightering operation, is critical to protecting both vessels. A reliable cargo transfer system, including well maintained hoses, is essential. Equipment standards for offshore lightering have been established by the industry (ITOL, 1990; ICS and OCIMF, 1997), but there are no guidelines specifically for inshore lightering. Additional guidelines and standards have been established by lightering companies, major tanker operators, and other industry entities.[1]

Lightering safety is also critically affected by external, environmental factors. These factors include weather and sea conditions, the location of oil pipelines buried in the ocean bottom (as factors in safe anchorages), and specific zones for lightering (or where lightering is prohibited) that have been established as regulatory and enforcement measures. In this chapter, physical systems, both on board and external to the vessel, are examined and opportunities identified for reducing the risk of accidents.

VESSELS AND SPECIAL SYSTEMS

Navigation systems, vessel maneuverability, and the control of vessels when approaching and departing the lightering area are all factors that affect safe operation. Electronic position-fixing by the differential global positioning system is used to locate operations very accurately. Vessels should also be equipped with accurate and up-to-date navigational charts to locate safe anchorages and to avoid hazards, fixed structures (which may not be well lit), and underwater obstructions (which are not always charted). Modern electronic chart displays can also be valuable navigation aids.[2]

Effective maneuvering controls are required on board both vessels during mooring and lightering at sea, as well as during unmooring operations. A few service vessels are equipped with bow thrusters, controllable pitch propellers, twin screws, and special rudders to aid in maneuvering. Most also have engine controls on the bridge and sometimes at other locations. Good systems for communications between key locations onboard each vessel and between vessels are critical.

The following sections examine some of the critical shipboard systems that could affect lightering safety and identify the ones that require attention to prevent future problems. The committee identified these systems based on its members' backgrounds and expertise in lightering as well as on a review of the literature and information gathered from industry, regulators, and others during the study.[3]

The key elements of vessels and special systems are vessel design, mooring systems, fendering systems, and hoses and transfer systems.[4] The key elements

---

[1]Companies that have in-house guidelines and standards include Chevron Shipping Company, Skaugen PetroTrans, Maritrans, SeaRiver Maritime, Conoco, Statoil, and Shell Oil Company.

[2]For a more detailed discussion of navigation aids and electronic traffic systems, see NRC 1994a, 1996.

[3]Information was gathered during meetings of the full committee and by individual members in Houston, Philadelphia, and San Francisco, as well as from discussions with various experts (see Appendix B).

[4]Navigation control and communications systems are also important, but concerns about their design, function, and effectiveness are not unique to lightering and therefore are not addressed in this report. These systems have been examined in previous studies by the National Research Council (1994a, 1996).

of the external environment are weather reports and warning systems, pipelines and other structures in the lightering areas, and lightering zones and prohibited areas.

Vessel Design

Lightering vessels may be engaged by a variety of operators and shipping companies. These vessels have a range of design features that make them more or less suitable for lightering activities. Vessels that have been built or converted for lightering, usually service vessels that are dedicated solely to lightering, have standard mooring, fendering, and hose transfer systems built in, and they engage in lightering on a regular basis.

Most of the firms involved in Gulf Coast lightering (and some on other coasts as well) use service vessels that are not dedicated to lightering. These vessels may split their time between lightering and other activities, such as short-haul deliveries of oil from Mexico or Venezuela to the U.S. Gulf of Mexico. Mooring lines, fenders, and hoses are usually delivered to these vessels just before a lightering operation, used during the offloading of one STBL, and then removed by the service company that was engaged to provide this equipment, expertise, and personnel for the lightering operation. These nondedicated vessels are of varying designs, so the special equipment must be able to accommodate a variety of on-board arrangements.

STBLs exhibit an even greater variety of designs and arrangements of equipment. An STBL can be any vessel from the world tanker fleet with several possible places of construction, any age, any flag of registration, and crew nationality, and so forth. General concerns about the quality of ships, crews, and operators engaged in oil transportation to U.S. ports (see NRC, 1994a) also apply to lightering but are not addressed in this report. The factors that affect lightering include the capabilities of the STBL and operating personnel.

Double-Hull Issues

Certain provisions of OPA 90 raise some safety issues that are unique to lightering. Under OPA 90, which requires that tankers calling on U.S. ports have double hulls, single-hull tankers of 5,000 gross tons or more will be excluded from U.S. waters as of 2010 (except vessels with double sides or double bottoms, which can be used until 2015). However, exemptions in this law will permit single-hull STBLs to lighter offshore in designated lightering zones until 2015. Thus, some single-hull tankers will continue to engage in lightering activities for five years or more beyond the date when they are no longer permitted to enter U.S. ports.

Another factor is a stability problem unique to certain double-hull tanker designs, even though they are built to meet current double-hull standards. A number of moderate-sized (under 160,000 DWT) double-hull tankers have been built without centerline bulkheads to divide very large cargo tanks. These "single-tank-across" tankers can have problems with intact stability[5] during loading and unloading of the large tanks (NRC, 1997). Single-tank-across tankers can become unstable and suddenly list to one side unless careful cargo loading and unloading practices are used (NRC, 1997). Several incidents involving the instability of double-hull tankers at U.S. and foreign terminals have been reported (NRC, 1997). It is not known whether these problems have ever arisen during lightering, but these incidents are cause for concern because more than half of the vessels of less than 160,000 DWT in the current fleet of double-hull tankers are single-tank-across vessels (NRC, 1997). The international tanker fleet is gradually converting to double hulls and will be composed almost entirely of double-hull tankers by 2023. It is conceivable that a vessel might suddenly list to one side during a lightering operation, roll against the other vessel, break the mooring lines, and break away. This situation is gradually being corrected as new tankers are built to new International Maritime Organization (IMO) standards that require inherent stability by design. Thus, the percentage of vessels with single-tank-across designs is expected to decrease each year.

The IMO, which is the United Nations agency responsible for maritime safety and protection of the marine environment, has addressed this issue in a draft circular that provides guidelines for ensuring intact stability during cargo transfer operations (IMO, 1997). A recent report by the National Research Council (1997) recommended that the USCG implement operational procedures and crew training for existing tankers that are subject to problems with intact stability and develop design requirements to ensure the intact stability of new double-hull tankers. Although the IMO and USCG efforts to address this problem appear to be adequate for shipping in general, lightering safety in particular will require that vessel operators and crews adhere strictly to the new guidelines and standards. Furthermore, as new tankers are designed, lightering safety is also likely to be improved by ensuring that mooring bitts and chocks, winches, line-handling systems, manifolds and cranes, and emergency quick-release systems all meet safety standards. This equipment is described later in this chapter.

Another aspect of tanker design that affects lightering is the much greater freeboard[6] of empty double-hull tankers as compared to single-hull tankers. Because of the large freeboard, the deck of a double-hull STBL averages about 80 feet above the waterline, roughly 60 feet above the deck of a service vessel, during a lightering operation. This results in an awkward, near-vertical arrangement of mooring lines that makes it difficult to hold the vessels together. To minimize this problem, companies have reduced the lightering weather window when

---

[5]Intact stability refers to the stability of an undamaged vessel.

[6] Freeboard is the distance from a vessel's waterline to the main deck.

extreme differences in freeboard exist. To optimize the effectiveness of the mooring lines, some operators limit the volumes of the last and next-to-last lifts from the STBL. Another recommended approach to the freeboard problem is to increase the allowed ballast capacity for the STBL to reduce freeboard and the vertical separation of the vessels. This approach would only be appropriate under unusual conditions and would probably require a modification of the current International Convention for the Prevention of Pollution from Ships (MARPOL, 1973), which defines exceptions to ballasting prohibitions.

Although no accidents have been reported from large freeboard differences between two ships, the committee considers that this is an area where risks of future accidents could be avoided by applying simple preventive measures. One reason no incidents have been reported to date could be that tankers with very high freeboards are just beginning to be commonly used in U.S. waters. The problem could grow in the future. Another reason could be that the skill and good judgment of lightering masters and others have prevented problems so far. However, committee members who have visited lightering operations and those who are experienced with lightering have pointed out that high freeboard is a problem worthy of attention. Preventive measures, such as ballasting, are straightforward and could lead to a safer operation.

The committee considered a number of approaches to solving the high freeboard problem, including enhancements to hose or mooring systems and design changes to double-hull ships, but concluded that these would either not be practical or would take too long in practice to implement. The most practical solution, and the one that had the best chance of being implemented, was to permit greater ballasting.

The committee recognizes that permitting greater ballasting of certain tankers under special circumstances could result in small additional discharges of oil entering the marine environment when the ship is deballasted. The committee did not compare the risk of allowing more dirty ballast on board with the avoided accident risk. But if a modification to MARPOL were adopted, it would be preceded by a full discussion of this issue by the IMO.

Engine Capabilities

When two vessels are brought together while under way, their speed affects the risk of a collision. The slower they are moving, the less likely they are to collide because of the angle of approach or some other factor. The greatest risk of a collision exists when two vessels of similar size traveling in excess of 5 knots are mooring.

Modern motor vessels cannot operate at very slow speeds, which can make both mooring and subsequent operations difficult, especially if a vessel is unable to anchor and the weather is inclement. At present, accommodating the limitations on the design and required operating range of modern diesel engines requires great skill and good judgment on the part of vessel operators. In the future, it may be possible to design slow-speed propulsion systems that are better suited to lightering operations.

Emergency Equipment

All vessels that engage in lightering operations within the EEZ must have a USCG-approved oil spill response plan, as required by OPA 90 (see Chapter 2, Box 2-1). Response plans are designed for a worst-case discharge and a substantial threat of a discharge of oil. The plans must specify the resources needed to respond to spills of various sizes in different environments and the contracting for a certain percentage of the needed resources. In addition, specific equipment must be carried on board. For example, tank vessels must carry spill-removal equipment, such as sorbents, hand scoops, shovels, buckets, portable pumps with hoses, containers, and protective clothing, in sufficient amounts to contain and remove on-deck oil cargo spills of at least 12 barrels.

Single-hull vessels that engage in lightering must also carry on-board equipment for emergency lightering transfer connections. The required equipment includes reducers, adapters, bolts, washers, nuts, and gaskets for at least two simultaneous transfer connections from the vessel's cargo manifold to cargo hoses of various sizes.

In addition, all tank vessels are required to have made arrangements for firefighting and towing. In some areas, operators are also required to arrange for access to additional emergency equipment. In San Francisco Bay, for example, standby vessels for both spill response and towing are required by state regulation. During deepwater operations at the LOOP, tugs carrying spill-response and firefighting equipment are always on scene during transfer operations. The standby tug can also assist a tanker that loses steering capabilities or has an engine failure. During offshore lightering, the availability and amount of firefighting and towing equipment on the scene, as opposed to on call, is usually up to the individual operator. Although standby equipment is seldom needed, prudent operators always provide it.

Mooring Equipment, Fenders, and Transfer Equipment

The three categories of special equipment necessary for the safe transfer of oil cargo between two vessels on the open ocean are listed below:

- a method of keeping the vessels together (i.e., a strong, well designed mooring system)
- a method of keeping the vessels apart and protecting them from each other (i.e., a fender system suitable for the individual operation)
- a reliable transfer system for moving the oil from one vessel to the other (i.e., hoses, connections, and equipment for connecting and disconnecting them)

The very steep angle of mooring lines, an unsatisfactory condition, is necessitated by large differences in deck heights from one vessel to another. Photo Credit: Chevron Shipping Co.

The key safety aspects of this equipment are discussed briefly in the following sections.

Mooring Systems

Conventional mooring systems are designed to secure a vessel at a dock, pier, or oil jetty. Over many years, the shipping industry has developed standard approaches and guidelines for the number, composition, and angles of mooring lines for mooring at a dock. These guidelines and standards do not, however, readily apply to lightering when the circumstances are very different, the forces and dynamics are much greater, and each lightering situation is unique. The ICS and OCIMF (1997) provided guidelines for mooring equipment used in ship-to-ship transfer operations. Many companies have also established their own standards.

Vessels moored together in an open seaway tend to move independently, and the motion can cause chafing and shock loads and the subsequent failure of mooring lines, particularly lines made of synthetic materials. Moreover, large changes in freeboard and extreme angles between the vessel leads can compromise the efficiency and strength of the lines. Therefore, a ship-to-ship mooring system designed for offshore use must be substantially stronger than a conventional mooring system. For example, the "tails" of mooring lines, which are usually made of nylon, should be rated at 125 percent of the strength of the lines to absorb shock. Short wire pennants can be attached to the tails to withstand abrasion. Vessels can also be fitted with roller chocks to ease the movement of mooring lines. Reliable chocks (or fairleads), preferably enclosed, can help control the lines. Roller chocks must be maintained properly or else they become useless. Well placed bitts on deck are also necessary to secure the mooring lines. The recommended mooring arrangement is shown in Figure 3-1.

In addition to being very strong, an offshore mooring system must also be designed to enable rapid breakaways, in case sudden changes in the weather warrant an emergency separation of the vessels. Too many mooring lines can impede breakaways without increasing the reliability or strength of the mooring system. The simplest and safest way to provide for rapid separation is to use quick-release connections.

The crews of both vessels should know, preferably in advance, that the mooring systems are adequate and compatible (i.e., the number of mooring bitts is sufficient and all lines are on permanent winch drums). This information is available from various sources, including the vessel owners and the OCIMF Ship Inspection Report (SIRE) program. The accessibility of SIRE records is discussed in Chapter 4.

The side-by-side mooring arrangement used in a typical lightering operation is only practical in low-to-moderate seas under reasonably good weather conditions. If the weather turns severe and waves reach a certain height, the operations

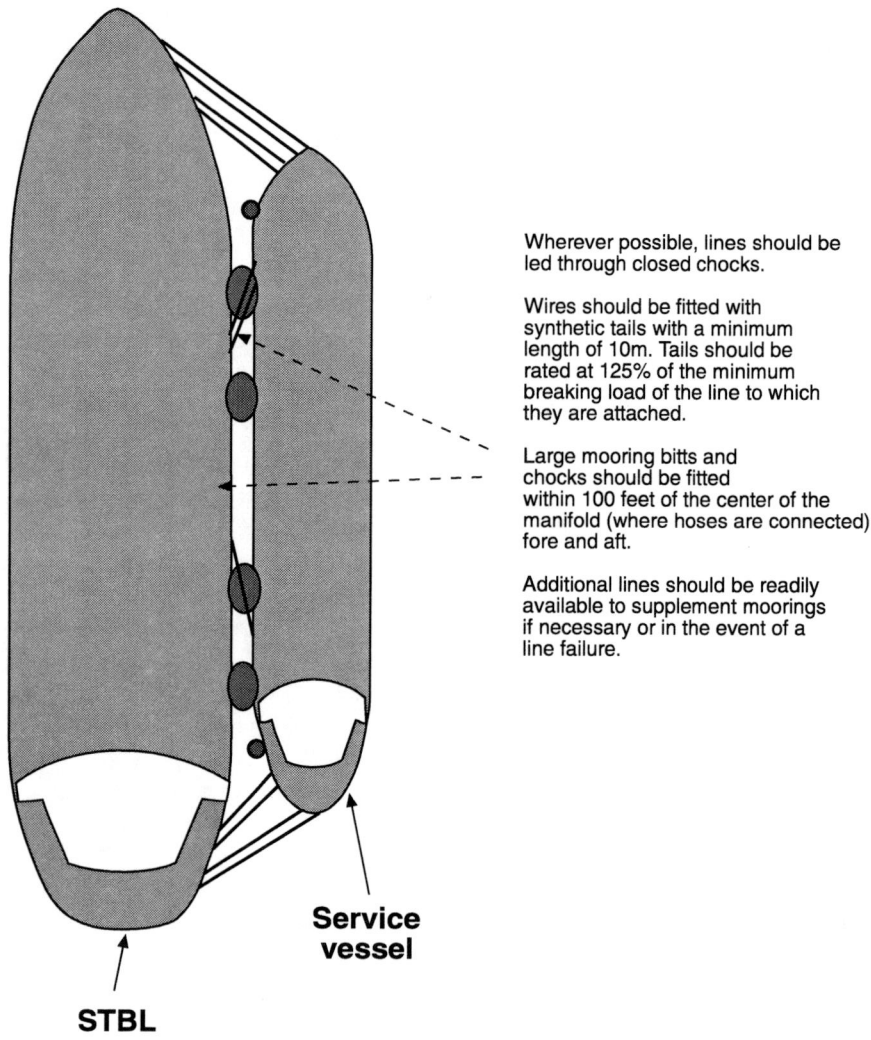

Wherever possible, lines should be led through closed chocks.

Wires should be fitted with synthetic tails with a minimum length of 10m. Tails should be rated at 125% of the minimum breaking load of the line to which they are attached.

Large mooring bitts and chocks should be fitted within 100 feet of the center of the manifold (where hoses are connected) fore and aft.

Additional lines should be readily available to supplement moorings if necessary or in the event of a line failure.

Service vessel

STBL

FIGURE 3-1 Recommended mooring arrangements for offshore lightering.

must be suspended and the vessels separated. Vigilance and good judgment on the part of all mariners are essential to avoid damaging either vessel.

Fenders

The OCIMF (ICS and OCIMF, 1997) and individual company standards specify the types and testing of fenders. Before a lightering operation begins, the

A workboat delivers fenders to a lightering operation. Photo Credit: Chevron Shipping Co.

mooring master or lightering service company representative tests the pressure of pneumatic fenders, which must be inflated according to the manufacturer's instructions (foam-filled fenders are sometimes used instead). Pneumatic fenders are most reliable when they are fitted with safety release valves to prevent them from bursting when compressed. In the past, some operators used truck tires instead of fenders in inshore waters. It is not known if contacts or accidents were caused by this practice, but by all accounts, this practice is no longer used.

Fenders are deployed between the STBL and the service vessel.

Primary fenders, which absorb the impact from the connection of the two vessels, must have the proper diameter in relation to the vessel's freeboard to prevent the fenders from riding up the sides of the vessel and rolling onto the deck. The OCIMF recommends that fender diameter not exceed half the freeboard at any time. The fenders should also provide maximum protection along the hull, allowing for both approach and departure angles. Experience and skill are required to place fenders properly because vessels meet at different angles. The speed of approach significantly affects fender requirements, and it is prudent to overestimate the approach speed when selecting fender size. The fender rigging system is also critical.

Secondary (or "baby") fenders provide additional protection in case the angle between the vessels exceeds the protective capability of the primary fenders. Secondary fenders are placed fore and aft of the primary fenders to prevent steel-to-steel contact. The flare and counter of the two vessels, aggravated by motion, results in an almost infinite number of possible contact points. Tankers have extensive parallel mid-bodies, where fenders work best. However, some modern vessels have short parallel mid-bodies, so secondary fenders must be available in appropriate locations. Fenders may have to be moved or adjusted to suit the situation. Some vessels that are routinely engaged in lightering are fitted with permanent mounting points to facilitate the handling of secondary fenders. Some operators of dedicated lightering barges have placed permanent molded-rubber fenders on the sides at these contact points, which eliminates the need to deploy fenders for each lightering operation.

The ICS and OCIMF (1997) guidelines cover the overall design and selection criteria for fenders, and operators refer to these guidelines in their internal company procedures. The guidelines specify numbers, types, sizes, and other criteria for fenders for particular operations and given conditions, such as vessel sizes and speed of approach. However, each operator should design a fendering system that suits the conditions of a particular lightering operation using the equipment performance data provided by the manufacturer. Most operators periodically test and inspect fenders and release valves. No specific testing and inspection standards have been established by regulation or industry guidelines, but prudent operators follow the same general practices they follow for all critical equipment.

Hoses

Many types of hoses are used in lightering. All of them must be certified to U.S. standards for this service. The transfer hoses have a thick outer rubber jacket with an inner liner for reinforcement. Hoses are generally connected to the manifold by means of a fully bolted flange, which provides a strong and reliable connection but is not conducive to rapid disconnection in emergencies. Experienced hose handlers can disconnect them quickly, but quick-release connections (e.g., KAM-locks) can also be installed. The releases must be USCG-approved quick-release mechanisms.

According to USCG MIN-MOD records (see Appendix C), there were nine hose ruptures during lightering operations in U.S. waters between 1984 and 1996; one additional rupture in 1998 is known to the committee. Although the numbers are small, ruptured hoses are more frequent than other causes of spills and, therefore, may present an opportunity for improvement. The committee's examination of federal regulations and industry standards for the construction, testing, and inspection of hoses revealed no obvious deficiencies, although there were some inconsistencies that should be reconciled.

According to the USCG regulations covering oil transfer hoses (33 CFR 154.500), the burst pressure must be at least four times the maximum allowable working pressure (MAWP).[7] The operator must clearly mark the hose with the MAWP. Industry guidelines for hose construction (OCIMF, 1978, 1995. Chevron, 1989) refer to a "rated pressure," which must be at least 225 psi but can be higher as specified by the purchaser. The rated pressure is supposed to account for dynamic loads from pressure surges, so it is likely to be somewhat higher than the MAWP.

The USCG testing requirements (33 CFR 156.170) specify a test pressure of one-and-one-half times the MAWP and a complete physical examination. The OCIMF guidelines require a test pressure equal to the rated pressure and some measurements of length that are not required by the USCG. The committee believes that operators should use the MAWP as a baseline for meeting USCG testing requirements, rather than mixing standards by testing hoses at one-and-one-half times the OCIMF rated pressure.

Vessel operators report that personnel involved in the day-to-day handling of hoses inspect them visually. Visual inspections are mandated in the DOI and are listed on OCIMF checklists (see Appendix H). Some companies retire hoses after only a few years in service, whereas others extend hose service life based on a review of performance and test data.

The industry guidelines for the construction and testing of hoses, together with the USCG regulations, provide high standards for the operations, testing, and inspection of hoses. Committee members who routinely conduct lightering operations noted that hose failures during testing are often caused by the separation of the internal liner. However, predicting these failures is difficult because many hoses are collapsible so the inner liners cannot be vacuum-tested for integrity. Prudent operators should adhere to the existing standards while remaining vigilant during cargo transfers to avoid unusual external chafing or axial loading, which can shorten the life of a hose or lead to a rupture. Hoses should be inspected regularly because they are exposed to more wear during lightering operations than when they are used at a terminal.

EXTERNAL FACTORS

Weather

Weather is a major risk factor in lightering. Weather is both a safety factor, because winds and sea state can affect vessel interactions, and a legal factor, because regulations specify the weather conditions under which lightering may take place in the designated lightering zones in the Gulf of Mexico (33 CFR 156.320). Thus, real-time information about the weather on the scene of the

---

[7]The MAWP must be more than the sum of the pressure of the relief-valve setting (or maximum pump pressure when no relief valve is fitted) plus the static head pressure of the transfer system.

lightering operation, at locations six to eight hours away, and as much as 24 hours in the future should be available.

As the two vessels come together, vessel operators use information about winds, currents, and sea state to the best advantage. This information also ensures that the vessels will not be "peeled apart" during the approach. Once the vessels are safely moored together, the weather is continuously monitored to ensure that conditions do not deteriorate to the point where safety is compromised. Sea state, winds, and currents also affect the unmooring of the two vessels.

If it appears that weather may become marginal, then the operators have several options. They may (1) terminate cargo operations, drain the hoses, and keep the vessels together until the bad weather passes; (2) terminate cargo operations, drain the hoses, and separate the vessels with the intention of coming back together later to finish transferring the cargo; or (3) continue operations. Interrupting an operation and dismantling connections takes time, so the longer the lead time the lower the risk associated with unmooring. When the weather is deteriorating, it is important to have accurate forecasts that give operators sufficient time to unmoor and move the vessels apart before the weather becomes too severe to accomplish this safely.

Mariners obtain offshore weather information from several sources. (Harbor and inland information is provided by separate systems, such as NOAA's physical oceanographic real-time system or private forecasting and broadcasting services, such as Navy Fleet Weather.) The National Data Buoy Center[8] operates offshore weather stations, funded by NOAA, that provide data to mariners by marine-band radio. Information from weather buoys and other sources is compiled by NOAA for weather updates broadcast by radio. The buoys and NOAA updates are the most heavily used, and probably the most accurate, sources of offshore weather information. Other sources of offshore weather data include the Texas Automated Buoy System, private weather forecasting and broadcasting services, and the U.S. Navy.

Anecdotal reports suggest that the available weather data sometimes do not meet the needs of lightering operators. The reasons include inappropriate locations of weather buoys, a lack of real-time information, and delays in repairs of weather buoys. For example, most weather buoys off the coast of California are located close to shore, and none of them provides useful real-time information. On the East Coast, one weather station (#44001) is located too far southwest to be useful for lightering near Delaware. And, during 1997, NOAA's National Weather Service (NWS) did not allocate funds for the maintenance and repair of some offshore weather stations. As a result, few real-time weather services were available in the Gulf of Mexico for at least six months. After funding from the

---

[8]Information about individual weather stations and environmental conditions in the area is available on the National Data Buoy Center home page (http://seaboard.ndbc.noaa.gov).

Minerals Management Service (MMS) for two weather stations (#42019 and #42020) off the Texas coast was depleted in June 1997, the NWS and NOAA were unable to secure funding for the redeployment of the weather buoys until November 1997. At another weather buoy (#42035), the wind direction and speed indicator were inoperable for most of 1997.

There appear to be a number of ways in which weather forecasting could be improved to support lightering operations. When contacted by the committee, the director of the NWS was unaware of the special needs of lightering operators but expressed a willingness to tailor weather forecasts to meet those needs. Better coordination among the agencies involved in shipping safety and weather forecasting could improve services that lightering operators rely on to maintain safety.

Charting Pipelines

More than 20,000 miles of petroleum pipelines can be found in U.S. coastal waters, mostly in the Gulf of Mexico (NRC, 1994b). Between 1967 and 1990, more than 95 percent of the pollution from pipelines was the result of maritime incidents, notably damage from anchors weighing up to 29 tons; four incidents accounted for 85 percent of the total pollution (NRC, 1994b). The best way to protect pipelines from anchor damage is to bury them in the ocean bottom, but shifting sediments in the Gulf of Mexico make it difficult to bury and maintain pipelines at adequate depths. With current technology, moving vessels cannot detect pipelines from far enough away to avoid them.

The vessels involved in lightering need accurate data on the location of pipelines to identify safe anchorages. ITOL has also identified the need for real-time mapping to track pipelines and platforms in the Gulf of Mexico (Caruselle, 1998). The voyage orders for an STBL usually recommend a specific area in which to anchor while the STBL waits for the service vessel. Lightering is done most efficiently and safely while the STBL is at anchor. Vessel masters often incorrectly assume that their voyage orders ensure them a safe, precise anchorage. In fact, to avoid pipelines, the master must have accurate, up-to-date notices, warnings, and charts of the pipelines in designated lightering zones as well as in traditional lightering areas. Warnings are provided in the North Sea, for example, where even temporary operations are publicized and where pipeline locations are disseminated as pipelines are installed.

As oil and gas exploration increases and new pipelines are installed in the Gulf of Mexico, the need for regularly updated data on the locations of pipelines is becoming more urgent. The problem is demonstrated by a pipeline map prepared for ITOL that shows very few safe anchorages in one of the designated lightering zones. In recent years, the MMS has had logistical problems collecting and updating this information. Data on pipeline installations are not collected on a regular basis, and the information takes too long to be recorded on charts or disseminated to users. The committee could not determine the extent of the backlog in collecting and disseminating pipeline locations from existing data. Although no pipeline ruptures from lightering have been reported, the proliferation of pipelines could increase the risk of spills in the future.

Governmental responsibility for oversight of these pipeline is shared by several agencies (NRC, 1994b). On the outer continental shelf, the Office of Pipeline Safety (OPS) of the U.S. Department of Transportation regulates approximately 13,000 miles of transmission lines (i.e., the larger, longer pipelines that carry oil and gas ashore), whereas the MMS, of the U.S. Department of the Interior, regulates approximately 4,000 miles of production pipelines associated with platform production systems. In state waters, the OPS regulates transmission lines, and the states regulate production pipelines. Under OPA 90, the MMS is responsible for oil spill prevention and response capabilities for all pipelines. The USCG is responsible for declaring pipeline hazards to navigation.

The most expedient approach to improving the quality and quantity of pipeline information available to mariners may be for the MMS, in cooperation with OPS and ITOL, to identify priority areas for immediate charting (ITOL has already requested this assistance [Caruselle, 1998]). The information should be updated at least every two years. Regulators might also consider establishing "pipeline-free" zones in both designated and traditional lightering areas, similar to the LOOP anchorage areas. These designations would assure mariners that specific areas are, and will continue to be, free of underwater obstructions and, therefore, provide safe anchorages.

Designated Lightering Zones and Prohibited Zones

To regulate certain lightering activities in the U.S. Gulf of Mexico, the USCG has established lightering zones for tankers retired under OPA 90, as well as other zones where lightering is prohibited. Both kinds of zones were established in 1996 following a rule-making process to implement the provisions of OPA 90. The designated lightering zones are for single-hull tankers that cannot enter U.S. ports under OPA 90 but can lighter offshore until 2015. The USCG has also established general regulations for lightering that apply to all vessels using the lightering zones. Prohibited zones (e.g., the Flower Garden Bank shown in Figure 2-2) were established to protect sensitive environmental areas.

So far, these zones are the only ones established for offshore lightering in the United States. All other lightering takes place in traditional lightering areas, which are used because the locations are convenient for operators and they do not create unsafe conditions. Notification of the appropriate COTP is required prior to any lightering operation in the EEZ involving cargo destined for a U.S. port, and all firms intending to lighter either offshore or inshore must request USCG approval before starting an operation. The USCG may require the use of a certain location or prohibit the use of an unsafe location.

Lightering operations in the Gulf of Mexico have led some observers to suggest that designated lightering zones should be established along the East Coast, or possibly off the coast of California. Increasing USCG control over lightering might provide some additional safeguards, but the current regulations require that designated lightering zones in the Gulf of Mexico be at least 60 miles offshore. Thus, as currently structured, the zone concept would not be appropriate for the east and west coasts, where inshore lightering is more common. On these coasts, other avenues of control are available through USCG authority based on COTP orders and not on national regulations. Any COTP can establish a lightering anchorage or regulated navigation area (see Table 2-3). Lightering anchorages have already been established at Big Stone Anchorage in Delaware Bay and Anchorage #9 in San Francisco Bay, for example. No safety problems have arisen in either of these busy locations.

For inshore lightering, the committee believes there is no overriding need for changing the traditional system of lightering in established anchorages. Some regions where lightering activity is new or growing, however, may require attention. The COTP in Long Island Sound, for example, plans to propose that the USCG establish regulated navigation areas under the district commander to enable officials to designate locations and minimum standards for the entire area. However, COTP authority to approve and monitor inshore lightering appears to ensure safety.

For offshore lightering (i.e., in international waters), the USCG's authority to regulate is somewhat uncertain outside of designated lightering zones. Nevertheless, the COTP's authority to require notification (and to regulate all vessels that eventually enter U.S. ports) has had the effect of setting guidelines for safe practices. The committee recognizes that establishing lightering zones offshore would have both positive and negative effects on safety. On the positive side, the USCG could set safety standards with enforceable regulations and could prevent operators with questionable standards from engaging in lightering. On the negative side, operators have noted that some safety problems are created by having to work within a designated zone that limits their flexibility on the high seas. For example, if a vessel drifts out of a zone during lightering, regulations could force it to unmoor before a lift is completed and moor again after reentering the zone. In this situation, the regulations could create hazards instead of preventing them.

The risks of environmental damage from lightering are related, in part, to the proximity of the operation to environmentally sensitive areas. In some regions, this problem can be addressed by prohibiting lightering in or near sensitive areas. The prohibited zones in the Gulf of Mexico are good examples. Some individuals have suggested that prohibited zones be established in other regions to lower risks under certain circumstances. In coastal waters and harbors, the USCG and other local authorities already have ample power to prohibit operations that could threaten sensitive environments. In international waters, however, U.S. jurisdiction and enforcement are limited. At present, lightering activity is substantial and widespread only in the Gulf of Mexico. The limited and familiar operations off the east and west coasts have been subject to reasonable oversight and review. Increased offshore lightering in the future could warrant a further evaluation of the benefits of designating formal prohibited zones.

The committee identified one apparent problem related to geographical constraints on lightering in the Gulf of Mexico. The traditional lightering areas sometimes become congested, and, as pipelines and platforms become more numerous in the Gulf, space is sometimes at a premium. ITOL has reported problems with access to the lightering areas off Southwest Pass, which is near the Ewing prohibited zone (Figure 2-2). The industry has asked if waivers could be requested from the local COTP to allow vessels engaged in lightering to drift through a prohibited zone when this is judged to be safer than turning or separating the vessels. The law currently prohibits certain vessels[9] from coming within 60 miles of shore, and the COTP cannot grant waivers except in an emergency. There is a formal procedure for applying for exemption at a higher level (33 CFR 156.110), but this procedure takes time. An extension of the COTP's authority to grant waivers in certain nonemergency situations may be a reasonable way to handle situations that could deteriorate and increase the risk of spills.

SUMMARY

Existing standards and guidelines have provided a solid foundation for lightering safety. However, the committee identified several aspects of vessel design, operations, and equipment that could be improved. First, lightering operations are safer when vessels have long parallel mid-bodies, an adequate number of well placed mounting points for fenders and enclosed chocks for mooring lines, and engines that enable controlled slow-speed operations. Second, all operators should use mooring lines with synthetic tails to absorb shock, adhere to appropriate standards when inspecting and testing hoses, and remain vigilant during their use. Third, the freeboard on very large STBLs should be limited. A comprehensive approach to this problem would be to work toward an international agreement allowing modifications to the ballasting system.

Regardless of whether and how these issues are addressed, industry standards and guidelines will continue to be voluntary. Based on the site visits by committee subgroups as well as the personal experience of various committee members, the committee recognizes that some vessels (especially STBLs) and equipment used in lightering will fail to meet the best standards and practices. For this reason, lightering operators must remain vigilant in reviewing the condition and characteristics of STBLs prior to lightering.

External conditions that affect the safety of lightering operations could be

---

[9]This provision applies to vessels retired under OPA 90 and vessels built without double hulls after the enactment of OPA 90.

improved. Gaps in the information available to mariners about local weather conditions on all coasts and the locations of oil pipelines in the Gulf of Mexico could be filled. The obvious remedy is for federal agencies to improve data collection and dissemination. A cost-effective approach might be for federal officials to meet with lightering companies and cooperative organizations to identify priorities and the best ways to meet them.

At present, the number of designated lightering zones or prohibited zones seems to be adequate, but one aspect of the current regulations could be improved. COTPs should have the authority to grant waivers allowing vessels engaged in lightering to depart from designated lightering zones when it would be safer than maneuvering or separating the vessels.

REFERENCES

Caruselle, P.A. 1998. Letter from Paul A. Caruselle, chairman of the Industry Taskforce on Offshore Lightering, to Chris Oynes, regional director GOM-OCS, Minerals Management Service, March 3, 1998.

Chevron Shipping Company. 1989. Oil Transfer Hose Testing—Gulf Coast Lightering. Internal guidance document used by Chevron, San Francisco. Unpublished.

ITOL (Industry Taskforce on Offshore Lightering). 1990. Industry Lightering Operations Supplement to OCIMF Ship to Ship Transfer Guidelines for U.S. Gulf Coast—Galveston Zone. Houston: ITOL.

ICS (International Chamber of Shipping) and OCIMF (Oil Companies International Marine Forum). 1997. Ship to Ship Transfer Guide (Petroleum). London: Witherby & Co., Ltd.

IMO (International Maritime Organization). 1997. Guidance on Intact Stability of Existing Tankers during Liquid Transfer Operations. Draft circular written by the IMO Marine Environmental Committee as part of ongoing regulatory development process.

NRC (National Research Council). 1994a. Minding the Helm: Marine Navigation and Piloting. Washington, D.C.: National Academy Press.

NRC. 1994b. Improving the Safety of Marine Pipelines. Washington, D.C.: National Academy Press.

NRC. 1996. Vessel Navigation and Traffic Services for Safe and Efficient Ports and Waterways, Interim Report. Washington, D.C.: National Academy Press.

NRC. 1997. Double-Hull Tanker Legislation: An Assessment of the Oil Pollution Act of 1990. Washington, D.C.: National Academy Press.

OCIMF (Oil Companies International Marine Forum). 1978. Hose Standards—Specification for Rubber, Reinforced, Smooth Bore, Oil Suction and Discharge Hoses for Offshore Moorings (Including Purchaser's Inspection Guide), 3rd ed. London: Witherby & Co., Ltd.

OCIMF. 1995. Guidelines for the Handling, Storage, Inspection, and Testing of Hoses in the Field, 2nd ed. London: Witherby & Co., Ltd.

4

Procedures, Practices, and Human Factors

The physical systems described in the preceding chapter are only part of the lightering picture. Additional safety considerations are related to operational procedures, industry practices, and human factors. A key operational procedure is communications, which includes fluency in English. Industry practices include international management and safety standards, cooperative systems to enhance learning and performance, and cargo identification and tracking. Relevant human factors include fatigue, training, and simulation. Risk evaluation and its proper application is another consideration.[1] This chapter examines opportunities for risk reduction in all of these areas.

OVERVIEW

With the assistance of ITOL and other participants at data-gathering meetings, the committee prepared Figure 4-1, a timeline and framework for planning and conducting lightering operations. This chart shows the steps that are usually taken from the initial planning of a voyage to import crude oil to the United States to the actual lightering operations and the departure of the vessel after it has discharged its cargo. The organizations typically involved include refineries, traders, ship operators, lightering service companies, regulators, and other oversight parties. Each step in the process is detailed in Box 4-1.

---

[1]Although a detailed risk analysis is outside the scope of this study, some of the methods that could be used are outlined briefly in this chapter.

Gulf of Mexico Ship-to-Ship Lightering
Voyage and Lightering Sequence of Events
Typical Time: Arabian Gulf to U.S. is 45 days

Task Name	Days Duration	1	15	30	45	
1. Set raw material requirements	1 day	—				
2. Purchase raw materials	1 day	—				
3. Determine STBL availability		◆				
4. STBL safety inspection	1 day	—				
5. Hire charter party	1 day	—				
6. Begin loading STBL		◆				
7. STBL en route to U.S.	23 days	————				
8. Send daily ETAs to charterer	23 days	————				
9. Send lightering plan to STBL	1 day	—				
10. STBL reviews plans	5 days	—				
11. Plan for each lightering operation	5 days		—			
12. Set lightering positions			◆			
13. Set lightering service details	7 days		—			
14. Set service vessel details	1 day		—			
15. STBL 72-hour ETA			◆			
16. Agent contacts USCG	3 days		—			
17. STBL arrives in lightering area			◆			
18. Schedule USCG oversight	1 day		—			
19. Service vessel arrives on site	1 day		—			
20. Purchaser gauging on board	1 day			—		
21. Obtain weather data	1 day			—		
22. Qualified individual arrives on site				◆		
23. Classification society arrives on site	3 days			—		
24. Verify checkoff lists				◆		
25. Lighter STBL	5 days			—		
26. Complete lightering of STBL					◆	

Legend:
STBL = ship to be lightered
Service vessel = ship receiving cargo

FIGURE 4-1 Timeline for Gulf of Mexico lightering.

BOX 4-1 The Lightering Process

The following is a detailed description of each step shown in the timeline (Figure 4-1).

(1) The lightering process is set in motion by a refinery's need for a certain type of crude oil, which dictates where the crude will be purchased. For example, a refinery designed to run a light, sweet crude would probably purchase it from the Arabian Gulf. The supply planners for the refinery project their needs 30 to 60 days in advance based on demand, seasonality, economics, and refinery capabilities.

(2) Once refinery plans are established, the crude traders, working under established agreements or in the open market, purchase the necessary "raw material barrels." The traders start with the projected dates when the crude is needed at the refinery and backtrack to a projected transportation date.

(3) Based on the purchases made by traders, marine groups secure the vessels needed to transport the crude. These vessels may be owned by the company or, more likely, chartered for a specific voyage.

(4) The STBLs that provide the economies of scale dictated by the purchaser are vetted (i.e., subjected to expert evaluation of their condition and operations) to ensure that they meet the minimum safety and operational criteria set by international law. The charterer may impose additional standards. These additional requirements are commonplace for STBLs making voyages to U.S. waters and may include: additional crew members, enhanced English-speaking capabilities, increased scrutiny of the vessel, review of the vessel's performance history (e.g., spills, equipment problems), the owner's performance history, and specific vessel designs. All acceptable vessels are then subject to negotiations to finalize the deal.

(5) The charter agreement between the parties stipulates the terms and conditions of the deal (e.g., rates, laydays, speed warranties, pumping warranties, insurance requirement, arbitration clauses).

(6) Once the charter has been signed, loading orders for the STBL are prepared by the charterer and provided to the vessel owner/operator. These orders include load port; laydays; grades, amounts, and specification of cargo to be loaded; discharge port and expected dates of arrival; stipulations for cargo surveyors; and cargo segregation instructions. The cargo is loaded onto the STBL according to the instructions. When loading is completed, the vessel owner/operator sends the charterer a report on the grades, amounts, and specifications of loaded cargo, as well as

(*continued on next page*)

BOX 4-1 *continued*

the estimated time of arrival (ETA) at the discharge location. A transit from the Arabian Gulf to the United States can take as long as 45 days.

(7) Once the STBL has departed from the load port, it generally maintains contact with the owner/operator, not the charterer. The owner/operator notifies the charterer of any unexpected delays, malfunctions, or incidents, which may or may not affect the delivery date of the cargo. Daily reports on ETA, vessel speed, and other information are usually available to suit the charterer's needs.

(8) When the STBL is 5 to 10 days away from the discharge location, the vessel owner/operator begins to provide more specific ETAs to the charterer. At this time, communications between the charterer and the STBL become more frequent.

(9) The charterer provides the STBL with the expected discharge sequence and lightering plan based on the refinery requirements and timing.

(10) If the lightering plan cannot be carried out as requested, then the STBL will provide information on its actual capabilities based on vessel stability and other factors. Information is passed back and forth until a lightering plan is agreed to by both the charterer and the STBL. The final plan is usually passed to the refinery, so that crude runs can be adjusted if necessary.

(11) Once the lightering plan is approved, a separate plan is developed for each lift, using much the same process as before. The charterer provides the names, cargo capacities, and ETAs of the service vessels.

(12) The STBL furnishes the recommended lightering positions based on whether it will use a designated lightering zone or a traditional lightering area.

(13) Communications now begin between the lightering service company and the STBL. The lightering service company provides information on the workboat, workboat ETA, mooring master, how the fenders and hoses will be deployed, and special instructions.

(14) The final schedule (including the number of lifts, service vessels, ETA, etc.) is provided to the STBL.

(15) 72 hours prior to arrival, the STBL provides its ETA to the USCG, charterer, lightering service company, STBL owner/operator, and charterer's agents.

(16) The STBL's agents, working with the owner/operators, determine if a TVEL (tank vessel examination letter) is needed and, if so, whether the vessel has a valid one. If the vessel has not been certified to international standards in the last 12 months, then an inspection must be scheduled with the USCG before lightering operations can begin.

BOX 4-1 *continued*

(17) The STBL arrives at the recommended lightering position.

(18) The STBL notifies the USCG of its position and expected lightering plan, including the number of lifts and the amount and grade of cargo per lift. If a TVEL is needed, then the inspection is carried out before any other operation can begin. The inspectors ensure that the vessel is in compliance with all applicable international requirements. Any other information needed by the USCG must also be provided at this time.

(19) Once the inspection is completed to the satisfaction of the USCG, the service company strings out the fenders and puts the cargo transfer hoses aboard the STBL (unless the service vessel is fully outfitted). The mooring master, who usually comes to the site on the workboat carrying the fenders and hoses, is placed aboard to inform the master, officers, and crew of the expected sequence of operations.

(20) An independent cargo surveyor boards the STBL and gauges the cargo tanks to ensure that there was no appreciable loss of cargo during the transit from the load port.

(21) Detailed weather information is provided to the workboat and mooring master, who will share the information with both the STBL and service vessel. Weather information is monitored continuously during the operation.

(22) The STBL owner/operator notifies the qualified individual for the vessel (required under OPA 90) that the vessel has arrived in U.S. waters and will be conducting lightering operations. The lightering plan is supplied to the qualified individual, who ensures that the requirements for the spill-response plan have been met.

(23) The STBL owner/operator ensures that the vessel classification society is available in case its services are needed. This is especially important if deficiencies are noted during the TVEL inspection, in which case the USCG will not reinspect the STBL but will depend on the classification society to verify that all deficiencies have been corrected.

(24) If the weather parameters are acceptable and the STBL has a valid TVEL, then the lightering begins.

(25) The lightering operation consists of the approach, transfer, and post-transfer phases. The approach phase encompasses the rendezvous in the lightering zone or area, the maneuvering of the STBL and service vessel alongside each other, the mooring process, and, if necessary, the anchoring of the STBL. The transfer phase includes the connecting hoses, the pumping cargo, and, if necessary, gauging cargo. The post-transfer phase includes the unmooring process and the departure of the service vessel.

The committee identified six general changes in the planning and operational process that would increase safety margins. The remainder of this chapter addresses these changes. In addition to reviewing current practices, the committee attempted to identify actions that could be taken to maintain the spill-prevention record in the future. The six areas of opportunity are listed below:

- vessel management and safety standards—identifying ways vessels and crews can maintain high standards of equipment safety and personnel performance through the use of international guidelines and codes, as well as through the training and certification of crews
- cooperative efforts to enhance learning and performance—using cooperative approaches, such as ITOL, harbor safety committees, or similar groups, to promote the sharing and discussion of critical safety issues among industry operators, regulators, government officials, and other service providers so that all parties can learn from each other and address concerns as they arise
- industry guidelines for lightering operations—using the ICS/OCIMF guidelines, which establish best practices for offshore lightering, to set standards for inshore lightering
- communications—improving the methods whereby the crew of each vessel communicates internally and with other vessels and how well they transmit vital operational information, especially during the critical mooring and unmooring steps
- human factors—preventing accidents through attention to human factors that could lead to fatigue and errors, and through the use of best training techniques
- risk evaluation—using formal analytical methods to identify operational risks and the possible causes of accidents

VESSEL MANAGEMENT AND SAFETY STANDARDS

Two sets of international shipping industry standards that have recently come into force will enhance, or at least maintain, the safety performance level of crews and operations. These management frameworks, which are mandatory and enforceable for vessels above a certain size threshold, are expected to promote improvements in operational performance and protect the marine environment. The standards will give the USCG further assurance that vessel quality is being maintained on all tankers. The USCG can exercise control measures (including detention of the vessel) if, among other things, a vessel crew's performance is judged to be unsatisfactory or does not meet requirements of the International Safety Management (ISM) code or Standards for Training, Certification, and Watchkeeping (STCW) certifications.

The ISM Code, which was adopted by the IMO in 1994, establishes detailed standards for the safe management and operation of ships and for pollution prevention. The code emphasizes management systems that promote best practices and procedures that rely on internal audits to protect the integrity of the system. To obtain certifications of compliance with ISM Code requirements, companies must undergo rigorous inspections and audits by either flag-state authorities or (when flag-state authorities so designate) by international vessel classification societies, such as the American Bureau of Shipping, Lloyd's Register of Shipping, Det Norske Veritas, or other members of the International Association of Classification Societies (IACS). The ISM requirements for tank vessels of 500 or more gross tons became mandatory in July 1998, the date when port-state authorities began checking to see that covered vessels had certifications on board. Informal industry surveys suggest that the vast majority of the world tanker fleet (unlike some other segments of the fleet) met the deadline.

The ISM Code does not apply to barges at all, or to tugboats under 500 gross tons. However, tugs and barges are covered by the Responsible Carrier Program (RCP), which was initiated in 1994 by the American Waterways Operators (AWO), the national trade association representing the U.S. domestic towboat, tugboat, and barge industry. The 300 member companies of AWO operate most of the towing equipment in the United States. Although the RCP is a voluntary program and not enforceable by law, the program establishes operating principles, practices, and guidelines that meet or exceed those currently required by federal law or USCG regulations. The philosophy guiding the program is that, although the federal government clearly has a role to play in ensuring safety and environmental protection by establishing a baseline, industry has the primary responsibility for ensuring safety. Unlike the ISM Code, which establishes a framework but does not address specific operational practices, the RCP specifies best practices and operational guidelines with which all members must comply (see Box 4-2).

The second management framework addresses crew competence and protection of the marine environment. In early 1997, new international standards for the skills and competence of seafarers entered into force. These standards were established by the 1995 amendments to the International Convention on STCW, which has been ratified by 120 countries representing more than 95 percent of the world's merchant fleet (IMO, 1998). These more stringent new standards and a better mechanism for ensuring accountability will help keep crew competence at a high level throughout the maritime industry, which will ultimately benefit lightering operations.

The changes are intended to increase the effectiveness of the 1978 convention, which was designed to establish consistent standards among maritime nations that previously had set their own standards for training, certification, and watchkeeping. By the late 1980s, it had become apparent that the convention's lack of precision had led to widely varying interpretations of the standards and

BOX 4-2 Responsible Carrier Program

The Responsible Carrier Program (RCP), a program instituted through the American Waterways Operators (AWO), is organized into three parts. The management and administration section requires a company to review eight principal aspects of its operations and develop written policies and procedures for each. The eight aspects are vessel operating procedures, safety policy and procedures, environmental policy and procedures, incident reporting procedures, emergency response procedures, internal audit and review procedures, organizational structure, and personnel policies. This section is designed to give a company the flexibility to tailor the program to meet its specific operational needs.

The second section is equipment and inspection. Minimum standards and requirements are set out for the vessel hull, machinery, firefighting and lifesaving equipment, navigation and communications equipment, rigging and towing gear, and environmental controls. These standards either reflect regulatory requirements or establish industry standards which must be reflected in the company's policies and procedures.

The third section deals with human factors, including manning, watchstanding and work hours, and training. This section outlines comprehensive criteria to be taken into account by companies in establishing safe manning levels for their vessels. It also establishes work-hour limits for all towing vessel personnel. Specific training is required on an initial and periodic basis depending on the individual's position; the requirements specifically cover captains, mates, engineers, deckhands, and tankermen. These requirements must be specified in the company's policies and procedures.

In the initial phase of the RCP, each member company's chief executive officer or senior marine officer was asked to notify the AWO president when the company had put the program in place. In 1997, the AWO Board of Directors approved a third-party audit requirement as a mechanism for making the RCP more effective and increasing its acceptance by governing bodies, industry customers, and the public at large. This requirement was approved for implementation in the first quarter of 1998, with the goal of having all members audited by January 1, 2000.

often ineffective administration and enforcement. As a result, STCW certificates did not provide reliable evidence of competence. The recent revisions define skills and competence in more detail; require direct control over and endorsement of the qualifications of masters, officers, and radio personnel; and make parties to the convention accountable to each other.

The revised convention deals with general provisions, master and deck department, engine department, radio communications and radio personnel, special training requirements for tankers and other types of vessels, emergency and safety procedures, and watchkeeping and fatigue. By August 1998, parties to the convention were required to inform the IMO of measures taken to ensure compliance, education and training, and certification procedures.

In recent years it has come to the attention of the international maritime community that certain flag states may not be upholding the signatory requirements of international treaties by allowing substandard vessels to operate and by licensing marginally trained officers and crews. In an effort to discourage these practices, many port states have stepped up inspections of vessels of suspect flag states. To ensure that all vessels comply with international standards, the IMO has declared that countries that are not signatories to the conventions must certify that they have in place a similar regime to ensure that their vessels meet the standards set forth by ISM and STCW. These vessels will be subject to the port-state and flag-state controls of countries in which they call, including the United States.

COOPERATIVE EFFORTS TO IMPROVE LEARNING AND PERFORMANCE

Lightering is a complex operation that demands foresight, experience, and accurate judgment. The vessel captains, mooring masters, officers, and crew all have to make decisions about complex weather conditions, variable crew skills, and different types of equipment so as to minimize the risk of accidents and spills. It is not possible to write specific, detailed regulations or procedures that cover every possible situation. Lightering operations can best be managed through a process that establishes threshold performance standards for equipment and personnel and, as long as these minimum standards are met, allows decision makers to exercise informed judgments in the context of specific events.

A model for this process is provided by ITOL, which has established a mechanism for the consensus-based development of industry standards for lightering in the Gulf of Mexico. Among its accomplishments, ITOL developed the *Industry Lightering Operations Supplement to OCIMF Ship to Ship Transfer Guide* (ITOL, 1990), which was approved by the local USCG COTP, Galveston, Texas, in 1990. In addition, ITOL has worked with the USCG to develop regulations for lightering zones and to develop pollution-response guidelines for the industry. ITOL also obtained approval from regulatory authorities for the use of oil dispersants in the lightering areas.

Through ITOL, representatives of federal and state agencies and industry can convene and cooperate to identify potential problems and establish procedures. ITOL has raised the awareness of local shipping agents who handle lightering activities and has encouraged all interested parties to express or respond to concerns about lightering operations, thereby fostering communication and creating an atmosphere of trust. The standards that have emerged from this process have established a high common denominator for safety.

The same organizational model might be used with equal success for lightering areas along the east and west coasts. Even if the number of participants is small, cooperation could be beneficial. This approach could be a vehicle for high-quality regional standards to become industry-wide standards without direct government regulation. Cooperative organizations could promote problem solving, interaction, and cooperation to enhance safety. Existing organizations, such as the AWO/USCG Safety Partnership,[2] could be vehicles for implementing this model.

An ongoing safety issue that might be resolved through cooperation is the need for detailed vessel information. All parties responsible for lightering safety need access to information about vessels. This is one of the most difficult issues facing the lightering community, according to ITOL, because much of the information is not readily available.

One source of information about tanker safety and inspections is the OCIMF, which has 34 member organizations that represent most of the world's major oil companies. The SIRE program is a voluntary reporting system that maintains computerized technical information about the condition and operational procedures of tankers. The database includes information from more than 17,000 inspection reports, with 600 new ones submitted each month. The primary objective of SIRE is to promote safety.

The information in the SIRE database is drawn from OCIMF member companies' in-house inspection programs and from comments by tanker operators. The standard format includes information about vessel safety and pollution prevention, certification, crew management, navigation, cargo handling, mooring, and engine room and steering gear. Under the current OCIMF charter, the SIRE information is available only to member oil companies, bulk oil terminal operators, companies that regularly charter tankers, and government agencies responsible for safety or pollution prevention. The information is not available directly to lightering companies.

It is not clear whether it would be legal for lightering companies to obtain SIRE information from traders who charter tankers or whether traders planning to lighter in U.S. waters could legally provide the information to lightering personnel. ITOL and other industry organizations could investigate this matter and propose a workable process. One approach might be to encourage revisions to the OCIMF charter to increase access to the database.

---

[2]The AWO/USCG Safety Partnership was established in 1995 to strengthen the working relationship between the barge and towing industry and the USCG and to provide a mechanism for cooperative action to achieve mutual goals (AWO/USCG Quality Action Team, 1997).

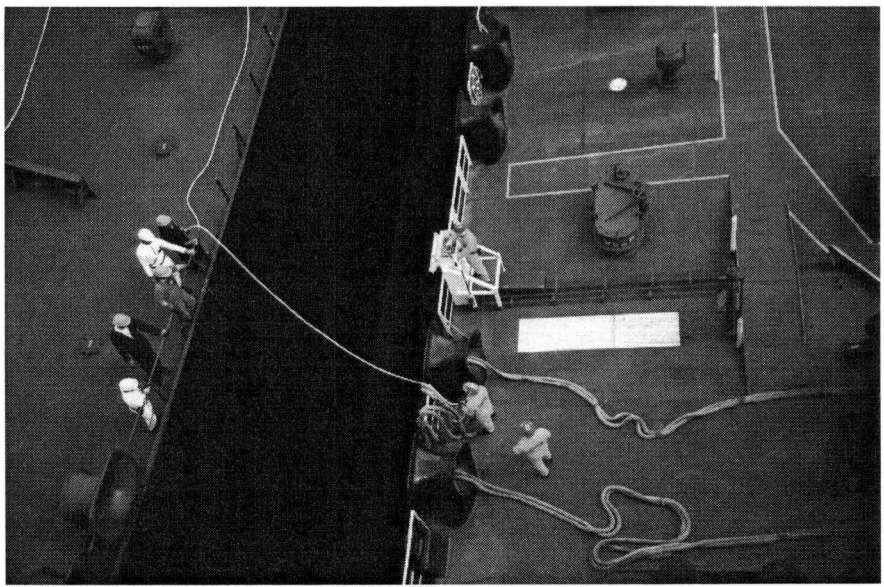

Crews aboard two vessels pass mooring lines.

INDUSTRY GUIDELINES FOR INSHORE LIGHTERING

Offshore lightering has more formal industry-developed guidelines than inshore lightering. Although USCG regulations and industry standards have been written for offshore operations, particularly those taking place in the Gulf of Mexico, no lightering guidelines have been written specifically for inshore operations. The AWO provides strong leadership for the tugs and barges that constitute much of the inland traffic on U.S. waterways, and there is no indication from the record of accidents and spills that standards for inshore operations are necessary. Nevertheless, given the fact that the operators and conditions in San Francisco Bay, Delaware Bay, and Long Island Sound differ from those in the Gulf of Mexico, establishing standards may be prudent. Standards could also establish a safety baseline if inshore lightering activity increases.

The OCIMF lightering guidelines (ICS and OCIMF, 1997) have established the international standards for offshore operations. Each edition of these guidelines, including a recent update, is developed by a committee of representatives of major operators. It may be advisable to use the same approach, or even the same standards, for inshore operations.

COMMUNICATIONS

The issue of maritime communications is being addressed by several organizations, including the American Petroleum Institute, the American Association of Port Authorities, and INTERTANKO. Although there is no need to duplicate their efforts or to address general maritime communications issues here,[3] lightering does pose unique challenges that warrant examination.

English is becoming the standard language in the international maritime industry, and English is required by most operators during lightering in the United States. Deck officers on international vessels are usually fluent in English because they are expected to communicate with other vessels by VHF radio in close traffic situations. Other crew members must also be fluent in English during lightering operations; the captains of both vessels and the mooring master must understand one another perfectly. Problems sometimes arise because of language differences or inconsistent terminology.

Every lightering operation is different. A few involve two U.S.-flag vessels with crews that are proficient in English. Most operations involve vessels with senior deck officers who are proficient in English but mixed crews representing many nationalities, some with minimal English skills. Still others may involve foreign-flag vessels with crews of the same nationality who converse in their native language and have limited English skills. Often the captain and the mooring master, lightering master, or assistant mooring master converse and exchange orders and commands in English and then translate them into other languages for the junior officers and crew. This may create problems because the mooring master, who usually speaks only English, cannot monitor the execution of orders.

Extra precautions must be taken when lightering operations involve crews that are not proficient in English. When completing the safety checklists or planning the operation, the mooring master or other individual in charge must identify communications barriers and take extra time to ensure that the plan is understood before initiating the operation. If the plan is not understood, then the operation has to be postponed until an interpreter can be present. If problems are identified early enough, perhaps in the pre-arrival message, an interpreter can be a condition of the lightering. Another approach, which is commonly practiced by major lightering companies, is to put an assistant lightering

---

[3]One general issue is the inadequacy of communications equipment. Most of the critical communication during lightering is done with hand-held VHF or UHF radios, which do not always function or interact well. Some companies use private channels, whereas others use standard marine channels, which are sometimes congested with other radio traffic. Other difficulties include battery or equipment failure, wind noise or interference, and background noise, such as engines and other conversations. Possible solutions to these problems, which are not unique to lightering, are outside the scope of this report.

master aboard the STBL to provide the necessary English language competency. This assistant can also carefully review lightering procedures with the crew prior to the operation to make sure everyone understands the procedures and commands that will be used.

HUMAN FACTORS

Human factors can both create problems and provide solutions. Approximately 80 percent of marine casualties have been attributed to human error (USCG, 1995; von Zharen, 1994). Human error has also been identified as a frequent factor in the small number of recent lightering-related spills. Human errors can be related to communication; job design; mental or physical fatigue; knowledge, training, or experience; or human-systems interfaces (von Zharen, 1994). Many maritime codes, standards, regulations, and studies address these issues, and there is no need to duplicate these efforts here. However, certain measures might be taken to help ensure that crews engaged in lightering are alert and well trained. Problems with communication were addressed in the previous section; this section addresses training, certification, and fatigue.

Training and Certification

The training and certification of all shipboard personnel engaged in a lightering operation are critical to safe operations. The lightering process is, however, only a part of any vessel's entire operating regime and must be considered as part of a much larger operating system. Recently, the international maritime industry has adopted comprehensive new standards known as the 1995 Amendments to the International Convention on STCW that represent a milestone for improving ship safety through enhancing personnel skills and training. These standards have come into force over the past year, and in August 1998, all parties were required to submit progress reports. The convention addresses minimum standards of competency, methods of demonstrating competency, and criteria for assessing compliance. It also addresses enforcement measures and specific port-state control measures. Some of the new requirements include mandatory rest periods, proficiency in English, basic safety training, and detailed descriptions of required knowledge, as well as methods for demonstrating competence for each, such as examinations, in-service experience, simulator training, and laboratory training, as appropriate.

The committee recognizes that improvements in training and certification within the international tanker fleet (represented by STCW) will have a significant impact on the safety of lightering practices because almost all vessels engaged in offshore lightering in the United States are foreign-flag vessels, and better general training will be key to the safety of all operations including lightering. The committee, however, has not conducted an extensive examination of seafarer training and certification issues, which would be beyond the scope of this study. The committee recognizes the importance of these issues, however, and the major changes that national and international authorities have recently put in place.

Most of the operational evolutions during lightering are not necessarily unique to lightering. For example, cargo transfer steps, such as pumping, loading, and hose handling, may be the same for lightering as for transfer steps at a buoy or a loading dock. The actual mooring of vessels can be as different from vessel to vessel as it is from dock to dock. One intent of STCW was to acknowledge that individual flag states could not certificate or create licenses for every marine job so it was made incumbent upon operators to ensure that vessel crews were qualified for the operations they were expected to perform. The committee considers that this approach is appropriate and will bring the best results.

Officers and crews on ships involved in lightering, both the STBL and service vessels, are qualified for tanker operations through licensing and training processes established by many international and flag-state requirements. Officer licensing standards provide for substantial training and time-in-grade experience. For example, a vessel master is usually a maritime school graduate (or has equivalent experience) with at least 10 years sailing experience and numerous training opportunities, including simulation. Most vessel officers are expert in mooring, unmooring, maneuvering, close navigation, cargo handling, communications, and crew supervision. Each flag authority is required to issue a certificate of endorsement stating that the seaman is qualified to serve on tankers. In addition, under STCW, officers and ratings assigned specific duties and responsibilities related to cargo must have a minimum service experience aboard a tanker and must complete a number of courses directly related to tanker operations. Among the tanker requirements of STCW are training for ship-to-ship-transfers of cargo to establish that the staff on board tankers is fully capable of handling the lightering operations. As an extra precaution, lightering companies assign mooring masters as well.

Lightering companies provide one or two persons expert in lightering and in local requirements to advise the ship staff during the operation. The mooring master is stationed on board the service vessel and, if there is a second lightering company staff member assigned to the operation, an assistant mooring master, he or she is stationed on the STBL to coordinate communications, ensure that everything is ready prior to commencement of the operation, and to assist during the operation. Lightering experts stress the need for skilled crews and experienced mooring masters. Thus, the technical "fix" for the human factors problem is to identify individuals who have good basic skills and then train them thoroughly. The government has not set standards or certifications for mooring masters, so each company now trains them to its own standards. Chevron, for example, certifies its lightering personnel according to its own standards, and the industry as a whole sets nearly equal standards.

A candidate for mooring master begins training by observing operations performed by a qualified mooring master, paying close attention to rendezvous positions, rigging methods, approach angles, mooring arrangements, the maintenance of lightering gear, fender retrieval methods, weather forecasting, and operational safety. The candidate also practices vessel separation and mooring under supervision. The training is tailored to each candidate's level of experience and usually lasts for several months. A review committee evaluates the candidate after training is completed. No written certification is issued, although candidates are approved in writing after a thorough evaluation at the fleet manager level. Usually at least four signatures are required before final approval. Chevron's West Coast operation has two levels of approval, one for summer and one for winter operations.

Some vessel operators and lightering companies use simulators for training candidates; others have adopted their own training standards. For example, Skaugen PetroTrans trains its mooring masters at the Ship Maneuvering Simulator Center in Trondheim, Norway. The company also uses a simulator for periodic instruction and testing of experienced personnel. Simulation has proven to be a valuable tool in aviation, and its application in maritime training was recommended in an earlier study (NRC, 1996). Simulation has also proven to be beneficial in training mooring masters. A growing body of literature indicates that simulation can help translate implicit knowledge into explicit knowledge and that it can facilitate the transfer of "experience" to other individuals quickly and precisely. The lightering industry understands these benefits and has applied them as needed.

Mooring masters are similar to maritime pilots. A pilot must be present on board every vessel entering state waters in the United States. Pilots are considered to be local experts and have the status of advisors but have no legal responsibility in the event of an accident, which is also true for mooring masters. The qualifications for mooring masters are high compared to the qualifications for most pilots. Lightering companies require a master's license for mooring masters and prefer that candidates have sailed as masters.

The committee found no evidence of a need for industry-wide training and certification programs specifically directed toward lightering personnel at this time. This is partly because training and certification is a more general issue applied to all shipboard personnel and partly because of the newly enacted STCW code. However, if lightering becomes more common and new companies enter the industry, then the need for certification should be investigated further.

Fatigue

The role of fatigue in maritime casualties has attracted a good deal of attention worldwide during the 1990s. The USCG accident database indicates that only 1 percent of all accidents are related to fatigue (Battelle Seattle Research Center, 1996), but this low rate may only reflect deficiencies in data collection.

Preliminary results of a Battelle study suggest that fatigue is a contributing factor in 16 percent of critical vessel casualties (Battelle Seattle Research Center, 1996).

The USCG is working on a multiyear project to develop better methods of identifying, investigating, and recording human factors in casualty investigations. Among the tangible results so far is a fatigue investigation work sheet, which is used by investigating officers to keep track of how long crew members worked and slept in the 24 hours preceding a casualty and to identify any other symptoms of fatigue (USCG, 1997).

Anecdotal reports suggest that crew members do not always get the rest stops required under OPA 90 (Section 4114), which states that crew members may not work more than 18 hours in any 24-hour period or more than 36 hours in any 72-hour period. Additional rest hours are prescribed by more recent requirements. Under STCW (Regulation VIII/1), watchkeepers must have a minimum of 10 hours of rest in any 24-hour period, and the hours of rest may be divided into no more than two periods, one of which must be at least 6 hours long. This provision is similar to the rest-hour requirement in 33 CFR 156.210(d) for operations in lightering zones. The mooring master is subject to restrictions on work hours if he or she is directing the movement of the vessel or assisting with navigation and cargo operations. Implementation of the ISM and STCW codes is expected to improve compliance with restrictions on work hours. Given the ongoing attention to this issue by the USCG, as well as officials in Canada and elsewhere, studies related specifically to lightering are probably not warranted.

RISK EVALUATION

Only a few formal risk analyses have focused specifically on lightering operations. For example, in a 1993 study of deepwater ports, the USCG concluded that lightering poses a substantially greater risk than using deepwater ports and a slightly greater risk than direct delivery of oil (USCG, 1993). However, that analysis may have inflated the risks of lightering because cargo transfers in the open ocean were considered together with transfers in port, which involve many small spills close to shore, increasing the risk of environmental damage.

Another analysis also noted that most lightering spills are small and concluded that the risks of lightering are manageable. "With adherence to the stringent federal and international requirements for prevention and response as well as the implementation of safety management mandates, the lightering process may be characterized as a relatively and predictably sound environmental risk" (von Zharen, 1994).

---

[4]The committee did not actually perform a risk assessment, primarily because it was outside the scope of the study and because the data available on lightering accidents and spills were not complete or reliable enough for a useful analysis.

The committee discussed the possible application of formal probabilistic risk assessment (or safety assessment) to help determine the risks posed by lightering operations.[4] A number of established assessment methods have been used in many other industries (e.g., nuclear power). Some methods, such as actuarial safety assessments, make extensive use of historical accident data and, therefore, may not be appropriate for lightering, which has seldom been implicated in past spills. Another method, known as engineering safety assessment, which requires an extremely detailed analysis of probabilities of failures for all systems and sub-systems, has also been used.

Another special type of assessment, the failure modes and effects assessment (FMEA), is nonquantitative and can be used as a first step in a formal process. The nuclear power industry has used this technique for many years, and the USCG has used it for studies of oil spill response plans. The FMEA method has proven to be very useful for evaluating operational risks and identifying the root causes of accidents. The process is designed to identify all possible ways a system can fail and all possible consequences of failures.[5] Consequences can be classified or ranked along a spectrum, from insignificant to catastrophic. This approach can be used as a management tool to trace a failure backward from critical events to determine the causative failure modes and take corrective action. This step-by-step formal process also takes into account expert opinion and system design.

FMEA analysis may be useful for analyzing lightering operations. Although it may not be necessary for operators with a long history of success, it might be useful for newer operations under special circumstances and for new lightering companies that want to learn from other's mistakes and successes. Companies and cooperative industry organizations, as well as USCG, may wish to become familiar with these techniques and to evaluate the possibility of using them in appropriate circumstances.

SUMMARY

The committee identified four possible safety improvements in current pro-cedures, practices, and communications related to lightering. The first two im-provements are related to industry cooperation. The benefits of information shar-ing and cooperative problem solving have been demonstrated by ITOL, which has contributed through published guidelines and meetings to the excellent safety record of the very busy lightering industry in the Gulf of Mexico in recent years. ITOL is also a model of a good working relationship between industry and the USCG. Similar cooperative forums on the east and west coasts, perhaps based on

---

[5]Examples of how to develop fault trees, which help determine the elements of a process or proce-dure that may contribute to a critical failure, may be found in Ang and Tang (1984). A detailed description of the procedure may be found in "Procedures for Performing a Failure Mode, Effects, and Criticality Analysis" (Mil-STD-1629A, 24 November 1980).

existing groups, could be established. Another improvement would be making reliable information about the condition of prospective STBLs available to lightering operators. This information is compiled in the SIRE system but is not currently accessible by lightering operators. Direct access to SIRE data could be provided only if OCIMF were willing to revise its charter regulations.

Another gap in the safety net is the absence of formal industry guidelines for inshore lightering. Although the committee did not identify any safety problems that reveal a need for guidelines, establishing a consistent safety threshold would be prudent. Many inshore operators use the OCIMF guidelines, which were written for offshore lightering, but some unique aspects of the vessels and practices used inshore may warrant the development of separate or modified guidelines.

Finally, the critical importance of good communications among the officers and crews aboard both vessels involved in a lightering operation underscores the necessity of all key personnel being fluent in English. Service vessel personnel should also be fluent in English.

REFERENCES

Ang, A. H-S., and W.H. Tang. 1984. Probability Concepts in Engineering Planning and Design. Vol. 2. Decision, Risk, and Reliability. New York: John Wiley & Sons.

AWO/USCG Quality Action Team. 1997. Tank Barge Transfer Spills: Managing Toward Zero Spills. Report presented to the AWO/Coast Guard National Quality Steering Committee. October 1997.

Battelle Seattle Research Center. 1996. Procedures for Investigating and Reporting Human Factors and Fatigue Contributions to Marine Casualties. Report No. CG-D-09-97. Available from the National Technical Information Service, 5285 Port Royal Road, Springfield, Virginia 22161.

ITOL (Industry Taskforce on Offshore Lightering). 1990. Industry Lightering Operations Supplement to OCIMF Ship to Ship Transfer Guidelines for U.S. Gulf Coast—Galveston Zone. Houston: ITOL.

ICS (International Chamber of Shipping) and OCIMF (Oil Companies International Marine Forum). 1997. Ship to Ship Transfer Guide (Petroleum). London: Witherby & Co., Ltd.

IMO (International Maritime Organization). 1998. IMO News 1(1): 6 pages.

NRC (National Research Council). 1996. Simulated Voyages: Using Simulation Technology to Train and License Mariners. Washington, D.C.: National Academy Press.

USCG (U.S. Coast Guard). 1993. Deepwater Port Study. Washington, D.C.: USCG Office of Marine Safety, Security, and Environmental Protection.

USCG. 1995. Prevention Through People Quality Action Team Report. Washington, D.C.: U.S. Department of Transportation.

USCG. 1997. Fatigue Investigation Worksheet. G-MOA Policy ltr 5-97, October 21, from the USCG Commandant.

von Zharen, W.M. 1994. Risk Evaluation of Ship-to-Ship Oil Transfer: An Assessment of Lightering as a Predictably Sound Environmental Risk: Inherent Relevant Concerns and Operational Safeguards. Galveston, Texas: Maritime and Environmental Management Research, Inc. and Texas A&M University Texas Institute of Oceanography.

5

Conclusions and Recommendations

Current lightering operations, which are conducted in a variety of locations in the United States using a variety of methods, are safe. The best available data on marine casualties and oil spills demonstrate that very few vessel accidents or spills in U.S. waters can be directly attributed to lightering operations. The safety record of lightering in recent years has been excellent. This observation is generally supported by representatives of oil, shipping, and lightering companies, as well as by government regulatory officials and members of environmental groups and the general public.

Lightering is an effective and efficient method of supplying U.S. refineries and storage facilities with foreign crude oil and petroleum products. The practice is less expensive than transporting oil by small tankers, minimizes deep-ocean traffic, and eliminates the need for very large oil tankers to enter U.S. ports. These advantages, combined with the good safety record of lightering, support the continued use of lightering.

The good safety record notwithstanding, the risk of spills from lightering could be reduced even further. The majority of the committee's recommendations are intended for industry. The rest are for the USCG and other federal agencies.

RECOMMENDATIONS FOR SHIPPING COMPANIES AND ORGANIZATIONS

Industry Guidelines for Inshore Lightering

The OCIMF *Ship to Ship Transfer Guide* provides vessel operators with minimum standards for safe offshore lightering operations. The guidelines were developed through a cooperative effort of the organization's international membership. Although the safety record of inshore lightering is excellent, the committee could not find universal guidelines defining minimum standards for inshore lightering. Many OCIMF guidelines can be readily applied to inshore operations, but this segment of the industry has a number of unique characteristics, such as the extensive use of barges and the frequent transport of specialized refined products, that may require the development of new standards. The committee, therefore, concludes that the industry would be well served by the development of standards and guidelines specifically for inshore lightering.

Recommendation 1. Industry organizations, such as the American Waterways Operators or cooperative organizations modeled on the Industry Taskforce on Offshore Lightering should develop standards and guidelines for inshore lightering operations. This document could either supplement or incorporate appropriate sections of the Oil Companies International Marine Forum guidelines for offshore operations.

Adherence to OCIMF guidelines

Judging from the data on lightering spills since the 1980s and the committee's site visits and review of the literature, lightering is a safe process that has a very low spill rate and manageable risks. Furthermore, the safety record of lightering is likely to stay the same or even improve as the overall quality in the shipping industry improves as a result of OPA 90, STCW, and ISM and as industry standards and practices for lightering specifically (such as the guidelines developed by OCIMF and ITOL) continue to evolve. However, anecdotal evidence and the experience and observations of committee members indicate that OCIMF guidelines are not applied uniformly throughout the shipping industry. Therefore, the committee concludes that, even though additional government regulations are not warranted, the industry should encourage all operators to adhere to best operational and management practices.

Recommendation 2. Chartering organizations should screen all prospective ships to be lightered to determine whether they meet Oil Companies International Marine Forum standards for vessels, equipment, and crews and should not charter vessels that do not meet these standards. As a supplementary measure to determine whether this self-policing process is effective, the U.S. Coast Guard should monitor the process and call for periodic reports when appropriate and needed.

Information on Vessel Conditions

The SIRE system is an oil industry initiative that facilitates the sharing of data among OCIMF members on vessel conditions. The data are used to make decisions about vessel chartering. The information in the SIRE database could also help lightering companies decide which vessels to charter and make plans to accommodate nonstandard features on particular vessels. Indeed, this information could help lightering companies deal with one of the major variables in lightering—the condition of STBLs. However, under current SIRE regulations, the OCIMF is not in a position to release vessel information to lightering companies. The committee concludes that the safety of lightering operations might be enhanced if lightering companies had access to the information in the SIRE system.

Recommendation 3. The Oil Companies International Marine Forum should consider making limited revisions to its Ship Inspection Report regulations to give lightering companies access to information on the condition of vessels.

Vessel Design, Construction, and Operation

The safety of lightering operations depends heavily on the compatibility of the two vessels involved and design features that support appropriate equipment. At a minimum, vessels must be designed and constructed to accommodate effective primary and secondary fenders and a strong, well balanced, flexible mooring system. Vessels must also be capable of maneuvering at low speeds for extended periods of time. A wide range of vessel designs, some more appropriate than others, are currently used for lightering vessels. Some attention has been paid to design issues of particular importance to lightering (e.g., the IMO guidelines for safe loading and unloading of single-tank-across double-hull vessels), and this positive trend should be continued. The committee concludes that attention should continue to be focused on vessel design, construction, and operation to support ship-to-ship transfer operations, particularly with respect to vessels that are likely to be used for lightering at some point in time. Issues that should be emphasized include the extent of vertical plating and parallel bodies on vessels, the size and placement of mounting points and lifting equipment, engine capabilities, and the potential for excessive freeboard.

Recommendation 4. To promote the adequate rigging of secondary fenders, the Oil Companies International Marine Forum should emphasize (e.g., in the next edition of its lightering guidelines) the need for vertical and flat surfaces as high as possible along vessel sides above the load waterline, with the maximum amount of vertical sides consistent with design requirements. In addition, mounting points, leads, and lifting equipment for secondary fenders should be positioned and sized for optimum effectiveness, and leads and securing facilities should be provided for primary fenders to ensure maximum coverage.

Recommendation 5. The Oil Companies International Marine Forum should focus on the need for vessels to have enough full-sized mooring bitts and enclosed chocks to secure the two vessels together a minimum of four lines forward and aft. A minimum of one full-sized mooring bitt and enclosed chock should be provided within 35 meters forward and aft of the manifold. All mooring lines should be secured by winches.

Recommendation 6. The Oil Companies International Marine Forum should focus on the need for vessels that are capable of slow steaming for extended periods of time (within the limited operating range of modern diesel engines) with fine control of engine revolutions to enable safe maneuvering during mooring and unmooring operations.

Recommendation 7. The Oil Companies International Marine Forum should recommend limited operating parameters for modern double-hull tankers used as ships to be lightered to accommodate excessive freeboard (up to about 85 feet) when the cargo tanks are empty, a condition that can degrade the integrity of the mooring between the ship to be lightered and the service vessel. At the same time, the International Maritime Organization should consider modifying MARPOL, Annex I, Regulation 13, or classifying lightering as an "exceptional case," to permit greater ballasting when transferring oil to a service vessel.

Lightering Equipment

A wide range of equipment is available for lightering purposes, and the specifications and arrangements outlined in the OCIMF guidelines are adequate. However, the observations and experience of committee members suggests that certain types of equipment are better for lightering operations than others. The shock-absorption capability of mooring lines is enhanced if the lines are fitted with synthetic tails. The use of truck tires instead of fenders, a practice that has been observed in some locations in the past, is questionable from a safety standpoint. Existing standards and guidelines for inspecting and testing hoses, especially the USCG's maximum allowable working pressure (as opposed to the OCIMF's "rated pressure") should be used as a baseline for testing hoses. The committee concludes that industry guidelines on the specifications and handling of equipment generally provide adequate margins of safety for lightering operations, but safety could still be improved.

Recommendation 8. Mooring lines should be fitted with synthetic tails and fenders designed specifically for lightering operations. Lightering operators should also adhere carefully to existing standards and guidelines with regard to the inspection and testing of hoses.

Communications

The safety of lightering depends heavily on effective communications between the STBL and the service vessel, as well as among the officers and crew members on each vessel. Difficulties sometimes arise because one or more key individuals on the STBL (which are usually foreign-flag vessels and often have crews of mixed nationalities) are not fluent in English. The committee, therefore, concludes that measures should be taken to ensure that key individuals are fluent in English.

Recommendation 9. Before initiating cargo transfer operations, the mooring master (or equivalent person in charge) aboard the service vessel should determine whether the key individuals on the ship to be lightered are fluent in English and can understand the lightering plans and respond to commands. If necessary, an individual (reporting to the lightering master or other official in charge of lightering) who is both fluent in English and knowledgeable about lightering should be put aboard the ship to be lightered prior to the transfer of cargo.

Cargo Gauging

After a lightering operation is completed, the cargo inspectors gauge the oil in the tanks of both vessels to determine the amount discharged and received. This gauging process can take more than two hours, and the measurements are often only approximate because the service vessel is moving in the seaway. In marginal, adverse, or worsening weather, the mooring master is anxious to separate the vessels. The cargo gauging process is repeated when the service vessel reaches port, and this figure is more accurate because the vessel is steady in port. The committee, therefore, concludes that cargo gauging offshore is redundant and, in marginal or adverse weather, constitutes an unnecessary risk, at least in U.S. waters.

Recommendation 10. To limit the time that vessels are alongside each other in a seaway and avoid delays in departure under adverse or marginal weather conditions, the Industry Taskforce on Offshore Lightering should suggest (e.g., in the next edition of its offshore lightering guidelines) that the mooring master and vessel master dispense with the inspector's gauging (at least on the service vessel) until the vessel is in port. If cargo quantities must be ascertained offshore, gauging should be limited to the ship to be lightered and should be done after the service vessel has departed. The cargo measurements for the service vessel could be telexed to the ship to be lightered.

RECOMMENDATIONS FOR THE U.S. COAST GUARD AND OTHER FEDERAL AGENCIES

Cooperative Problem Solving

ITOL is a cooperative organization, established at the suggestion of the USCG, that promotes self-policing in the industry and, in partnership with the USCG, promotes continuous improvement in the lightering process in the Gulf of Mexico. Among its accomplishments, ITOL has published the *Industry Lightering Operations Supplement to OCIMF Ship to Ship Transfer Guide*, which was approved by the USCG in 1990. In addition, ITOL has worked with the

USCG to develop the regulations for designated lightering zones and to write pollution-response guidelines for the industry. ITOL has also obtained pre-approval for the use of oil dispersants in traditional lightering areas.

The effectiveness of ITOL is reflected in the low incidence of oil spills in relation to the quantity of cargo transferred in the Gulf of Mexico. The committee concludes that the organization is an excellent example of how industry and government cooperation can reduce the risks associated with oil spills.

Recommendation 11. The U.S. Coast Guard should encourage the lightering industry on the east and west coasts to adopt or adapt the Industry Taskforce on Offshore Lightering model as part of their program to promote problem solving, interaction, and cooperation to enhance safety and environmental protection. Cooperative arrangements could be initiated through existing mechanisms, such as the American Waterways Operators/U.S. Coast Guard Safety Partnership.

Weather Forecasting

Accurate, timely weather forecasts are essential to safe lightering operations. Forecasts are necessary for operations in designated lightering zones and traditional lightering areas on all three U.S. coasts. The committee found a number of problems with respect to the availability and usefulness of marine weather forecasts for lightering purposes. Reported problems include the inappropriate location of weather buoys, a lack of real-time information, and delays in repairs to weather buoys. The committee concludes that the safety of lightering operations would be enhanced if weather information was more reliable and accessible.

Recommendation 12. The U.S. Coast Guard, in consultation with the lightering industry, should work with the National Weather Service and the U.S. Navy to select appropriate locations for weather buoys and to tailor weather data and forecasts to support operations in both designated lightering zones and traditional lightering areas. The National Weather Service should take on this task as a priority to improve the delivery of weather information to enhance safety in offshore operations.

Waivers for Departures from Designated Lightering Zones

Vessels that must use designated lightering zones are barred by law from departing from these zones during lightering operations, except in emergency situations when waivers are granted by the local COTP. The shipping industry has attempted to modify this restriction, arguing that unforeseen circumstances sometimes extend the duration of a lightering operation to the point that the vessels approach the zone boundary while they are still moored together and under way. To avoid crossing the boundary, the vessels must either be maneuvered while they are moored together or separated prior to completing a lift. Either option poses safety risks. The committee concludes that this maneuvering or separation could unnecessarily increase the risk of spills.

Recommendation 13. The U.S. Coast Guard captain of the port should be given the authority, based on a case-by-case review of individual requests and circumstances, to allow vessels to leave designated lightering zones for safety reasons while still engaged in lightering.

Collection of Data on Spills

An immense amount of information is available on maritime accidents from the USCG databases, state agencies, and private sources. However, it is extremely difficult to sort through this information to gather reliable data on the history of oil spills related to lightering in U.S. waters. The difficulties include inconsistent reporting and ambiguous information on the underlying causes of accidents. To create a national picture of the lightering-related spill pattern, accident data of varying degrees of detail and reliability must be combined from various sources. The committee gathered sufficient anecdotal and experiential evidence to verify the very low rate of lightering-related spills, but the process was laborious. The committee, therefore, concludes that the analysis—and presumably the prevention—of accidents would be enhanced by the development of an accurate, comprehensive database on maritime oil spills that would enable users to sort all spills in U.S. waters by the causes of accidents, including equipment failure modes, the activities (e.g., lightering) under way at the time, and other pertinent variables.

Recommendation 14. The U.S. Coast Guard should develop, or hire a private contractor to develop, an accurate, comprehensive computer database on maritime oil spills that can be searched and sorted by pertinent variables, including the causes of accidents.

Human Error

Human error is a factor in a large percentage of maritime casualties, including the few lightering-related spills. Although human factors is an important issue in improving maritime safety, many other studies, as well as existing and emerging standards and regulations, are already addressing this subject. Moreover, problems in this area are not unique to lightering. The committee concludes, therefore, that human factors in maritime safety are likely to be addressed adequately in other studies and in the development of improved standards and practices in the maritime industry in general and do not require special attention with respect to lightering. However, lightering companies and operators should continue to be involved with industry improvements.

Charting Pipelines

Vessels engaged in lightering in the Gulf of Mexico may have to anchor, either while awaiting the arrival of a service vessel or during the cargo transfer process. Anchors can weigh as much as 29 tons—more than enough to damage, and cause a spill from, one of the growing number of oil pipelines on the ocean bottom. To avoid damaging pipelines and causing spills, the operators of STBLs and service vessels need accurate data on the location of underwater pipelines in designated lightering zones and traditional lightering areas. Federal agencies do not currently collect and publish these data on a regular basis. A recent private survey highlighted the need for this data. The survey revealed that the bottom in one designated lightering zone is covered by pipelines. The committee concludes that there is an urgent need for the regular collection of accurate data on the locations of pipelines in the Gulf of Mexico, especially in designated lightering zones and traditional lightering areas.

Recommendation 15. The Minerals Management Service (of the U.S. Department of the Interior) and the Office of Pipeline Safety (of the U.S. Department of Transportation) should develop and implement a plan to collect and compile accurate data on the location of pipelines in the Gulf of Mexico and make the information available to the operators of vessels that engage in lightering. Priority should be placed on data collection in designated lightering zones and traditional lightering areas, and the data should be verified and updated on a regular basis.

Recommendation 16. To ensure safe anchorages amid the increasing oil and gas exploration activity, the U.S. Coast Guard should seek authority to designate "pipeline-free areas" where new pipelines cannot be laid.

Acronyms

AWO   American Waterways Operators

COFR   certificate of financial responsibility
COTP   captain of the port

DOI   declaration of inspection
DWT   deadweight tons

EEZ   Exclusive Economic Zone
ETA   estimated time of arrival

FMEA   failure modes and effects assessment

IMO   International Maritime Organization
ISM   International Safety Management Code
ITOL   Industry Taskforce on Offshore Lightering

LOOP   Louisiana Offshore Oil Port

MAWP   maximum allowable working pressure
MIN-MOD   Marine Investigation Module
MMS   Minerals Management Service

NRC   National Research Council

NWS   National Weather Service

OCIMF   Oil Companies International Marine Forum
OPA 90   Oil Pollution Act of 1990
OPS   Office of Pipeline Safety

RCP   Responsible Carrier Program

SIRE   Ship Inspection Report Program
STBL   ship to be lightered
STCW   International Convention on Standards of Training,
                 Certification, and Watchkeeping for Seafarers

TVEL   tank vessel examination letter

ULCC   ultralarge crude carrier
USCG   U.S. Coast Guard

VLCC   very large crude carrier

Appendixes

Appendix A

BIOGRAPHIES OF COMMITTEE MEMBERS

DON E. KASH (*chair*) holds the John T. Hazel, Sr., and Ruth D. Hazel Chair of Public Policy at the Institute of Public Policy, George Mason University (GMU). He is also a professor in the GMU Department of Public Affairs. His fields of research include science technology and public policy, energy policy, and policy analysis. Dr. Kash was a research professor of political science at the University of Oklahoma for more than 20 years and held similar positions at Indiana University, Purdue University, and the University of Missouri. He has also held several government management positions, including chief of the Conservation Division and assistant director for regulation at the U.S. Geological Survey. Dr. Kash has extensive experience chairing or serving as a member of committees for the congressional Office of Technology Assessment and the National Research Council (NRC) as well as other government advisory organizations. He is a past member of the NRC's Marine Board. He has published numerous books and articles on subjects related to science and engineering and their effect on public policy. Dr. Kash has a Ph.D. in political science from the University of Iowa.

TRICIA CLARK is the maritime affairs coordinator for the Oil Spill Division of the Texas General Land Office, which develops and enforces state regulations affecting the maritime industry. She serves as a liaison between the state and the maritime community and directs safety task forces, regional response teams, safety advisory committees, and other initiatives. Previously, Ms. Clark was a licensed deck officer for ARCO Marine and has extensive knowledge of tanker operations, fleet management, and lightering. She was also the Texas state liaison to the U.S. Coast Guard (USCG) team that developed regulations to implement the Oil Pollution Act of 1990. In that position, she contributed to the negotiated rulemaking process for oil spill response plans and the USCG's *Deepwater Ports Study.* She has a B.S. in marine biology from Texas A&M University and a Chief Mate unlimited tonnage license from the USCG.

ALFRED COLE is a master mariner and lightering master for Chevron Shipping Company in Pascagoula, Mississippi. He has worked for Chevron for 23 years. As a lightering master in the Gulf of Mexico for more than 10 years, he has supervised approximately 600 lightering operations. He was the principal technical advisor in the planning, organization, and implementation of Chevron's 1996 project to evaluate the use of lightering systems in the open ocean off Southern California. He also has served as master of very large crude carriers operating worldwide, director of training and development, and terminal manager in Australia. Prior to his career with Chevron, he sailed for 13 years with the Royal Fleet Auxiliary in the United Kingdom. Captain Cole chaired the Oil Companies International Marine Forum task force that developed the most recent industry guidelines for offshore lightering operations.

EDWARD C. CROSS is president of Plimsoll Shipping, Inc., a marine surveying company in Houston. He previously spent 27 years with Mobil Shipping, where he served in various deck officer positions, including master with specialty in lightering. He initiated, and for 10 years supervised, Mobil's lightering operations in the Gulf of Mexico. He then became the safety officer for Mobil's U.S. and international fleet. After leaving Mobil, he was the operations manager for a lightering company working on the East Coast and in the Gulf of Mexico. Now, in addition to running his own surveying company, he is a consultant to several major lightering companies on safety and pollution prevention matters and occasionally works as an independent mooring master for these and other firms. He has a master mariner's degree from the University of Bristol, United Kingdom.

DUANE H. LAIBLE is president of The Glosten Associates, Inc., of Seattle, a marine engineering and naval architecture consulting firm. He has extensive experience in the design and construction of a variety of vessels, including tugboats, ferries, specialty barges, hydrofoils, catamarans, and research vessels. He has supervised major design projects and managed construction and conversion programs. Mr. Laible's other research activities include simulations of ship maneuvering operations and safety assessments of ships and harbors engaged in oil transportation. His firm has conducted assessments of tanker escort regulations in the San Francisco Bay, Alaska, and Puget Sound regions. He has a B.S. degree in naval architecture from Webb Institute and attended the University of Washington Graduate School of Business Management.

J. BRADFORD MOONEY, JR., *NAE,* is a consultant in ocean engineering and research management to universities and industry. He is a U.S. Navy rear admiral, retired, and a former president of Harbor Branch Oceanographic Institution. He has broad experience in management, research, education, training, and other areas. His Navy career included assignments as chief of naval research, oceanographer of the Navy, and naval deputy to the National Oceanic and Atmospheric Administration. He has extensive experience with submarines and deep-submergence vehicles, having served as pilot of the *Trieste II* in the successful 1964 search for the sunken Navy submarine, *Thresher*, and was founder of the Navy's first fleet operational deep submergence command. Admiral Mooney is a former member of the Marine Board and has served as chairman or a member of various National Research Council study panels. He received a B.S. degree from the U.S. Naval Academy and pursued postgraduate management studies at George Washington and Harvard universities.

STEPHEN D. RICKS is president of Clean Bay, Inc., in San Francisco. He manages the oil spill cleanup cooperative's activities, which include contingency planning, training, purchasing and maintenance of equipment, and spill response. Previously, he was vice president for Pacific Refining Company, where he was responsible for all refinery operations, including safety and pollution prevention programs. Mr. Ricks also has extensive experience with other firms in refinery operations, including oil storage and transportation terminal operations. He has a B.S. degree in chemical engineering from the University of California, Davis, and served in the U.S. Air Force.

EDWIN J. ROLAND is president of Bona Shipping (U.S.), Inc., a tanker operating company in Houston. He has extensive experience in the oil transportation business, having previously served as vice president of operations, planning, and transportation for Amoco Oil Company; president of Amoco Transport Company; vice president of Holland America Line; vice president of Coastal Corporation; and vice president of Conoco Shipping Company. Prior to that, he served 11 years in the U.S. Coast Guard. Mr. Roland is a member of the American Bureau of Shipping, Lloyd's American Committee, Webb Institute Board of Trustees, and boards of the U.S. Chamber of Shipping and Liberian Shipowner's Council. He has a B.S. degree from the U.S. Coast Guard Academy, an M.S. degree in nuclear engineering and naval architecture from the University of Michigan, and an M.B.A. degree from Iona College.

RICHARD J. STEADY is manager of regulatory affairs and compliance for Maritrans Operating Partners, L.P., a major petroleum transport company in the U.S. coastal trade. He has worked for Maritrans in various management capacities. During his service with the firm's Tampa operations, the company conducted offshore lightering near Galveston, Texas. Recently he has been directly responsible for a fleet of lightering vessels operating in the Delaware Bay and has directed the lightering coordinator team responsible for ensuring safety throughout the cargo transfer process. Mr. Steady is now responsible for monitoring state and federal rules that affect lightering vessels. He also serves on the regional risk assessment team addressing many issues related to tank vessel operations in the Northeast and on a variety of U.S. Coast Guard task forces in the Philadelphia area. He has a B.S. degree in mathematics and mechanical engineering from the University of New Hampshire and an M.B.A. degree from Temple University.

JOHN B. TORGAN is the Narragansett Bay Keeper with Save the Bay in Providence, Rhode Island. He leads the organization's program to protect the environmental integrity of the bay and its tributaries through sampling, research, and education. He develops outreach activities and other communications programs to bring problems to the attention of the public. He has also performed research on wildlife habitats in the region and provided testimony on ecological issues. Prior to his current position, Mr. Torgan performed ecological research and field studies in New York and Michigan, as well as fishery studies in rivers near hydroelectric dams. He has a B.S. degree in environmental studies and biology from Union College.

W. M. VON ZHAREN is an associate professor of environmental law and admiralty law at Texas A&M University. She is also maritime policy and management coordinator at Texas Institute of Oceanography and a member of the graduate faculty in the Department of Oceanography. She is widely published in a number of fields, including environmental management systems and stewardship of marine resources. She is currently editing a textbook on the waterborne transportation of hazardous chemicals. She has also published many articles on environmental risk management and recently conducted a risk assessment of offshore lightering activities in the Gulf of Mexico. Dr. von Zharen is counsel for environmental affairs for the American Bureau of Shipping's Marine Services, Inc. She is chair of the American Bar Association's Marine Resources Committee, a member of the Houston-Galveston Navigation Safety Committee, a member of the Industry Taskforce on Offshore Lightering, and a proctor in the Maritime Law Institute. Dr. von Zharen, who has J.D. and L.L.M. degrees in international law, was previously an attorney for the Exxon Shipping Company. She was also a Fulbright scholar and has studied in Denmark, Sweden, and Germany.

Appendix B

COMMITTEE MEETINGS, SUBGROUP MEETINGS, AND SITE VISITS

First Committee Meeting
August 5–6, 1997, Washington, D.C.

An overview of the U.S. Coast Guard's role in U.S. lightering activities and spill prevention, its regulatory authority and actions, and its expectations of the Marine Board study
LCDR Stephen L. Kantz, U.S. Coast Guard

Perspectives on current and future U.S. lightering activities and oil spill risks
Joe Cox, U.S. Chamber of Shipping
Dennis Bryant, Haight, Gardner, Holland, and Knight
Jonathan Benner, INTERTANKO
George D. Pence, Louisiana Offshore Oil Port

Second Committee Meeting
October 2–3, 1997, Houston, Texas

Lightering practices in the Gulf of Mexico were discussed by the following members of the Industry Task Force on Offshore Lightering (ITOL):
Paul Caruselle, SeaRiver Maritime and ITOL chairman
Ray Ambrose, American Eagle Tankers
Richard Ford, Aramco Services
Bob Carson, OMI Petrolink
Trygve Munthe, Skaugen PetroTrans
Michael A. Curtis, Skaugen PetroTrans
Don Prouty, ITOL

U.S. Coast Guard operations, regulatory practices, and experience related to lightering in the Gulf of Mexico
Captain Kevin Eldridge, USCG Captain of the Port (COTP) on
LCDR Gregory Buie, USCG Marine Safety Unit, Galveston

Perspectives on Gulf of Mexico lightering practices and opportunities to reduce risks of accidents
  Robert T. Bush, Neptune Marine Consulting, Inc.

Estimates and projections of current and future U.S. lightering patterns and volumes as related to crude oil imports and sources
  William R. Finger, ProxPro, Inc.

**Visits to Lightering Operations
by Subgroups of the Committee**

August 28, 1997
Visit to Gulf of Mexico lightering of cargo from 530,000-ton tanker into 80,000-ton service vessel approximately 60 miles off the Texas coast (arranged by SeaRiver Maritime).

October 1, 1997
Visit to Gulf of Mexico lightering of cargo from 150,000-ton tanker into 80,000-ton service vessel approximately 60 miles off the Texas coast (arranged by Skaugen PetroTrans).

October 30, 1997
Visit to offshore Gulf of Mexico lightering of cargo from 300,000-ton tanker into 80,000-ton service vessel at Pascagoula, Mississippi, lightering area (arranged by Chevron Transport Co.).

November 12, 1997.
Visit to Delaware Bay lightering of cargo from crude oil tanker into tug-barge unit (the service vessel) at Big Stone Anchorage in Delaware Bay (arranged by Maritrans, Inc.).

**Third Committee Meeting
November 13–14, 1997, Philadelphia, Pennsylvania**

Overview of lightering practices and the oil spill prevention record in the Delaware Bay region
  Michael Nesbitt, Maritrans Operating Partners, L.P.

U.S. Coast Guard history and current activities related to East Coast lightering and regulatory oversight
  Captain John Veentjer, USCG COTP, Philadelphia
  Captain Peter Mitchell, USCG COTP, Long Island Sound

Environmental Organizations' perspectives on lightering operations in the Delaware Bay and concerns about oil spill risks
    Jerry Shields, Greenwatch Institute
    Grace Pierce-Beck, Delaware Audubon Society

American Waterways Operators' perspectives on inshore harbor lightering on the East Coast
    Herb Walling, Moran Towing Co.

Committee Subgroup Meeting
January 15–16, 1998, San Francisco, California

Environmental concerns regarding current and planned lightering operations in the San Francisco Bay and offshore California region
    Suzanne Rogalin, California Coastal Commission
    Joan Lundstrum, San Francisco Bay Commission

U.S. Coast Guard and state of California oversight and regulatory practices related to West Coast lightering operations and plans for the future
    Captain Harlan Henderson, USCG COTP, San Francisco
    CDR Jim Watson, Marine Safety Office, San Diego
    Peter Bontadelli, California Office of Spill Prevention and Response

Current industry operations and practices and future plans related to inshore and offshore lightering on the U.S. West Coast
    Dennis R. Rement, Chevron Shipping Co.
    Richard A. Smith, SeaRiver Maritime, Inc.

Fourth Committee Meeting
March 19–20, 1998, Irvine, California

Overview of U.S. Coast Guard contracted study of automatic shut-off valves for preventing oil spills
    LCDR Stephen Kantz, USCG

Appendix C

U.S. Coast Guard Data on Lightering Incidents, 1984 to 1996

TABLE C-1 U.S. Coast Guard CASMAIN Database of Lightering Incidents in U.S. Waters, 1984 to 1996

Location	Vessel	Gallons	Primary Cause	Secondary Cause
EAST COAST				
Baybridge Anchorage, NY Harbor, 1993	Hiltra	200	unknown	unknown
NY Harbor Upper Bay, 1988	Cibro Albany	126	equipment failure	hose rupture
NY Harbor Upper Bay, 1988	Jarama	63,000	intended discharge	ballast pumping
Arthur Kill, Linden, NJ, 1988	Jarama	3,000	unintended discharge	ballast pumping
Bigstone Anchorage, Delaware Bay, 1994	Protank Medway	126	unknown	unknown
North Atlantic, Delaware, 1986	Jahre Pearl	42	unintended discharge	tank overflow
Bigstone Anchorage, Delaware Bay, 1986	Hera	1	equipment failure	valve failure
Bigstone Anchorage, Delaware Bay, 1986	Interstate 138	3	equipment failure	valve failure
Slaughter Beach, Delaware Bay, 1991	Interstate 52	1	structural failure	hull rupture, leak
Limestone Bay, St. Croix, 1988	Berge Princess	2	unintended discharge	NEC
WEST COAST				
Long Beach Harbor, Port of LA/LB, 1986	Prince William Sound	2	equipment failure	hose rupture
Port of Richmond, San Francisco Bay, 1986	Barge 1	42	tank spill	tank overflow
Long Beach Harbor, Oahu, Hawaii, 1989	Exxon North Slope	1	uintended discharge	NEC
Honolulu Harbor, Oahu, Hawaii, 1993	Titan Express	50	unknown	unknown
Prince William Sound, Valdez, AK, 1993	Fort Liscum	1	unknown	unknown
GULF COAST				
Navigable Waters, Lake Charles, LA, 1987	DF 101	1	unintended discharge	NEC
Navigable Waters, Texas City, TX, 1991	S-2011	1	structural failure	hull rupture, leak
SW Pass Lightering, Gulf of Mexico, 1994	Kraka	1	unknown	unknown
Gulf of Mexico 12-200, 1990	Energy Progress	2	equipment failure	gasket failure
Intercoastal Waterway, Burns, LA, 1994	7008	2	unknown	unknown
Gulf of Mexico 12-200, 1990	Solena	3	unintended discharge	NEC
Galveston Bay, Galveston, TX, 1987	Alison C	5	unintended discharge	NEC
Gulf of Mexico 12-200, 1989	White Sea	5	tank spill	tank overflow

TABLE C-1 continued

Location	Vessel	Gallons	Primary Cause	Secondary Cause
Lower Mississippi River, New Orleans, 1990	Hollywood 1513	5	structural failure	hull rupture, leak
LOOP Terminal, Gulf of Mexico 12-200, 1992	Berge Nisa	5	unknown	unknown
Navigable Waters, Amelia, TX, 1993	Cape Charles	5	unknown	unknown
Gulf of Mexico, Galveston, TX, 1988	Loire	10	unknown	NEC
Gulf of Mexico Coastal, 1988	Chevron Stream	10	equipment failure	NEC
Mississippi Sound, 1988	Chevron Stream	10	unintended discharge	NEC
Gulf of Mexico 12-200, 1990	Hansa Vega	10	unknown	NEC
Navigable Waters, Texas City, TX, 1991	H.T. Co. 1802	10	structural failure	hull rupture, leak
Navigable Waters, Lake Charles, LA, 1991	Navidad	15	unintended discharge	tank overflow
Gulf of Mexico 12-200, 1989	Esso Bermuda	16	unknown	NEC
Gulf of Mexico 12-200, 1989	Texaco Wilmington	20	unknown	NEC
Gulf of Mexico 12-200, 1990	Alandia Breeze	20	equipment failure	pipeline rupture
Gulf of Mexico 12-200, 1994	Venilza	20	unknown	unknown
Lower Mississippi River, Davant, LA, 1990	Trade Quest	40	equipment failure	NEC
Gulf of Mexico, Galveston, TX, 1995	Hellspont Grand	40	unknown	unknown
Gulf of Mexico Coastal, 1988	Samuel H. Armacost	42	equipment failure	NEC
Gulf of Mexico 12-200, 1990	Iseult	42	structural failure	hull rupture, leak
Gulf of Mexico 12-200, 1990	Esso Nassau	42	structural failure	ballast pumping
Gulf of Mexico 12-200, 1991	Nor Explorer	50	equipment failure	hose rupture
Gulf of Mexico, Pascagoula, MS, 1993	Chevron South America	63	unknown	unknown
Gulf of Mexico 12-200, 1989	Esso Mexico	84	equipment failure	hose rupture
Gulf of Mexico, Freeport, TX, 1986	Artemis Garofaldis	100	equipment failure	hose rupture
Gulf of Mexico, Galveston, TX, 1988	Katrine Maersk	126	equipment failure	hose rupture
Gulf of Mexico 12-200, 1989	Stena Explorer	126	equipment failure	hose rupture

Mobile River, Mobile, AL, 1990	Hollywood 2304	126	structural failure	NEC
Intercoastal Waterway, Bourg, LA, 1994	Lebeof Tower LA	126	unknown	unknown
Gulf of Mexico, Corpus Christi, TX, 1993	Front Leader	168	unknown	unknown
Navigable Waters, Sulphur, LA, 1986	Amazon Venture	210	unknown	NEC
Gulf of Mexico, 1990	Alandia Breeze	210	unintended discharge	valve failure
Lower Mississippi River, Marrero, LA, 1991	Hollywood 1062	252	NEC	NEC
Gulf of Mexico, Galveston, TX, 1986	Red Sea	420	structural failure	tank rupture
Gulf of Mexico 12-200, 1990	British Respect	420	unintended discharge	tank overflow
Gulf of Mexico 12-200, 1990	Burmpac Bahamas	500	unintended discharge	NEC
Gulf of Mexico 12-200, 1996	Chaumont	630	unknown	unknown
Intercoastal Waterway, High Island, TX, 1986	Hollywood 2004	840	structural failure	tank rupture
Gulf of Mexico, Port Aransas, 1989	Knock Nalling	1,000	equipment failure	valve failure
Gulf of Mexico, Freeport, TX, 1986	Kate Maersk	5,250	structural failure	hull rupture
Gulf of Mexico 12-200, 1989	Knock Taggert	21,000	structural failure	hull rupture

NOTE: NEC = not elsewhere categorized, thus the cause is unknown

Appendix D

DATA ON EAST COAST LIGHTERING OPERATIONS AND INCIDENTS

1) **Delaware Bay Lightering-Big Stone Anchorage (from Maritrans Corp.)**

1993–1997

Approximately 100 million barrels per year were lightered at Big Stone Anchorage from 1993 to 1997. No pollution incidents were reported as a direct result of the lightering process as defined in this study.

From 1985 to 1992, only one incident occurred (1992) which resulted in the release of approximately 5 gallons of crude oil. The spill was the result of a loose butterworth plate on the deck of the vessel.

2) **Long Island Sound Lightering-Various Locations** (from USCG COTP, November 1997)

During calendar year 1997, approximately 10 lighterings totaling 0.2 million barrels occurred in the Bridgeport lightering area, and nine lighterings totaling 1.2 million barrels occurred in the New Haven lightering area.

No spill incidents associated with lightering in Long Island Sound were reported during 1997 or during 1993 to 1997.

3) **East Coast Offshore Lightering**

No reliable statistical data are available at this time, but the volume of oil lightered appears to be quite small. No spill incidents have been reported in recent years.

Appendix E

DATA ON WEST COAST LIGHTERING OPERATIONS AND INCIDENTS

TABLE E-1 Summary of Oil Spills, California, 1992 to 1997

Year	Terminals		Vessels[a]	
	Incidents	Quantity (gallons)	Incidents	Quantity (gallons)
1992	18	1,777	18	1,256
1993	20	1,088	16	479
1994	16	507	17	679
1995	20	228	28	3,158
1996	12	339	12	286
1997	9	163	9	461
Totals	95	4,102	100	6,319
Attributable to:				
Pipelines	11	3,009	V/L Ops.	4,266

[a]34 incidents occurred during bunkering and accounted for 1,783 gallons. The largest spills were 1,000 gallons during the bunkering of a warship and 1,974 gallons during the loading of a barge.
Source: California State Lands Commission

TABLE E-2 Summary of Oil Spills, Washington State, 1992 to 1997

No spills on record during lighterings.
Recent spill data indicates that approximately:
- 75 percent was from vessels
- 18 percent was from shore facilities
- 7 percent was from pipelines

Causes:
- 56 percent organization/management deficiencies
- 20 percent equipment failures
- 24 percent human error

There are no indications or records of any spills as a result of lighterings

Source: Washington Department of Ecology

TABLE E-3 Chevron, Summary of Lightering, West Coast

		Lightering	Quantity of Oil	Spills (gallons)
Offshore				
San Clemente	1970–1979	250	75,000,000 Bbls.	0
Pacific Area Lightering	1996–1997	22	40,000,000 Bbls.	0
Inshore				
L.A. Harbor	1980 to date	70	17,000,000 Bbls.	0
San Francisco Bay	1992–1997	57	8,000,000 Bbls.	0
Totals		399	140,000,000 Bbls.	0

Source: Chevron Shipping Co.

TABLE E-4 British Petroleum, Summary of Lightering, West Coast, 1987–1997

	Lightering	Quantity of Oil	Spills (gallons)
Long Beach Harbor	98	34,855,585 Bbls.	5
San Francisco Bay	14	1,420,004 Bbls.	0
Puget Sound	159	27,241,360 Bbls.	0
Totals	271	63,000,000 Bbls.	0

Note: One spill (not related to lightering) was caused by a fracture in the bottom hull plate of a barge.
Source: British Petroleum

TABLE E-5 Exxon, Summary of Lightering, West Coast, San Francisco Harbor, 1992–1997

Year	Lightering	Quantity of Oil	Spills (gallons)
1991	187	40,765,767 Bbls.	0
1992	190	44,341,669 Bbls.	0
1993	187	42,298,141 Bbls.	0
1994	182	42,752,591 Bbls.	0
1995	189	44,825,615 Bbls.	0
1996	178	43,251,976 Bbls.	0
1997	103	23,269,520 Bbls.	0
Totals	1216	281,505,279 Bbls.	0

Source: Exxon

TABLE E-6 Summary of West Coast Lightering

	Lightering	Quantity of Oil
Exxon	1,212	281,500,000 Bbls.
British Petroleum	271	63,000,000 Bbls.
Chevron	399	140,000,000 Bbls.
Totals	1,882	484,500,000 Bbls.

Note: Total of 5 gallons was spilled due to a hull crack in a barge.

TABLE E-7 Chevron's Overall Lightering Experience, 1970 to 1997

	Barrels Lightered	Lighterings	Spills (barrels)
Pascagoula	1,700,000,000	4,550	10
Other U.S. Gulf	120,000,000	360	
Pacific Area Lightering (1996-1997)	30,000,000	32	
Other West Coast	100,000,000	380	
Totals	1,950,000,000	5,322	10

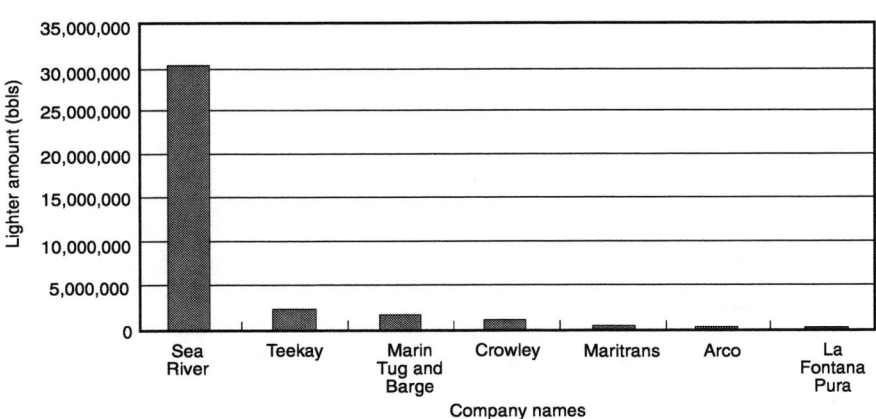

FIGURE E-1 San Francisco Bay lightering, 1997.

Appendix F

State of California—The Resources Agency
Department of Fish and Game
1416 Ninth Street
P.O. Box 944209
Sacramento, CA 94244-2090

June 27, 1996

Mr. Steven Hillyard
Manager, Government and
   Public Affairs Division
Chevron Shipping Company
555 Market Street
San Francisco, California 94105-2870

Dear Mr. Hillyard:

Thank you for providing us with a comprehensive briefing concerning the "Pacific Ocean Lightering Plan" earlier this month. Now that the first evolution is nearly completed and we have the initial reports from the United States Coast Guard (USCG), it would appear that the safety issues we discussed at the meeting are being addressed, are working, and will result in a safe operation during the six-month evaluation period.

I am sending this letter to follow up on several items agreed to at the meeting and telephone calls between your office and the Office of Oil Spill Prevention and Response (OSPR) as well as to confirm that we will be receiving copies of the documents that you are forwarding to the USCG.

As mentioned at the briefing, we are looking forward to reviewing the information obtained during the six-month evaluation and working with Chevron and the USCG on the long-term approach to the lightering program if it proves to be successful. We are mindful of your desire to implement the Pacific Ocean Lightering Plan as proposed and then develop a long-range plan based upon its results and lessons learned. We are prepared to work with Chevron Shipping to attain our mutual goal of continued smooth, safe transfers.

If lightering becomes standard practice, it is our understanding that the Middle East crude oil that is imported to southern California waters would be a substitute for Alaska North Slope crude oil which would otherwise be shipped in coastwise transits to the final destination terminals in Los Angeles and San Francisco. Thus, this lightering results in less overall exposure of crude oil to California's coastal resources. Coupled with the use of double-hull tankers traveling the final segment into California ports, as opposed to the older, single-hull Jones Act vessel used in the current Trans-Alaska Pipeline System (TAPS) trade, we view this operation as a net expansion of the margin of safety for transportation of crude oil into California ports.

As also discussed and agreed, for those vessels which would move lightered crude along the coast north to San Francisco, their routing would be in compliance with the voluntary 50-mile standoff routing that is observed by the TAPS trade today. That is, vessels carrying crude would remain at least 50 miles offshore until they intersect with traditional harbor approach lanes. We would like to see those routes depicted as well as those for the shuttle lighterer serving southern California ports.

It is our understanding that Chevron plans to have support vessels on scene during transfer operations, and has contracted for two vessels from certified oil spill response organizations to be available if a discharge occurs. This is consistent with the plans that Chevron has on file with our office. We are looking forward to a written specific response plan for the lightering program. We realize that the operation is being done in international waters and does not require a USCG permit (per our communication with the 11th District) and that no formal filing is due to the State. Therefore, we appreciate your willingness to provide us with the information sought.

In a related matter, we agree that the use of dispersant may be an appropriate response option in the event of a spill and note that the lightering is taking place within the Regional Response Team (RRT) "quick approval zone". As covered in the meeting at our office, we would like you to consider the use of Corexit 9500 as opposed to dispersant currently available in the West Coast response community. We have reviewed the fate and effect characteristics of Corexit 9500 and find that its application would be more suitable given the characteristics of the shipped product and the environment. Please advise us if you choose to accept this recommendation. In the event that use of such a response option appears viable, we will assist in the expedited approval of their application. As we committed at the meeting we are prepared to share all of the information we have on the biota in the lightering area to help enable us to make timely, informed decisions should the unlikely need arise.

At the meeting we provided information to you about commercial and recreational fishing interests which would likely be operating in the area of the lightering and proposed that you contact them directly. If you need any further information after you have talked to them please let us know.

We are appreciative of the fact that the lightering operation now under evaluation is being done in an even more conservative operational manner than current USCG lightering requirements when it comes to factors such as wave height, wind, and weather conditions. In the event that this operation matures into a routine practice in the years ahead, we would like to be provided with an operation manual, or operation plan for the lightering operation with written procedures in place stating under what conditions transfer operations will be suspended, weather, sea state, etc. We are sensitive to the proprietary nature of this information and of course would treat it accordingly.

Thank you for your cooperativeness in this important matter. If you have any questions or are in need of additional details concerning the information requested above, please contact Mr. Carl Moore, of my staff, who can be reached at telephone number (916) 327-9938 or at the letterhead address provided above.

[original signed by C.F.R.]

C. F. Raysbrook
Deputy Administrator
Office of Oil Spill Prevention
and Response

CFR.mld

Appendix G

LIGHTERING ZONE REGULATIONS

[These Regulations (excerpted from the Code of Federal Regulations) have been promulgated and are enforced by the U.S. Coast Guard. They enact specific provisions of the Oil Pollution Act of 1990 (OPA-90)]

Requirements from 33 CFR Part 156 Subpart C:

156.320 Maximum operating conditions.

Unless otherwise specified, the maximum operating conditions in this section apply to tank vessels operating within the lightering zones designated in this subpart.

(a) A tank vessel shall not attempt to moor alongside another vessel when either of the following conditions exist:
(1) The wind velocity is 56 km/hr (30 knots) or more; or
(2) The wave height is 3 meters (10 feet) or more.

(b) Cargo transfer operations shall cease and transfer hoses shall be drained when:
(1) The wind velocity exceeds 82 km/hr (44 knots); or
(2) The wave heights exceed 5 meters (16 feet).

156.330 Operations.

(a) Unless otherwise specified in the subpart, or when otherwise authorized by the cognizant Captain of the Port (COTP) or District Commander, the master of a vessel lightering in a zone designated in the subpart shall ensure that all officers and appropriate members of the crew are familiar with the guidelines in

117

paragraphs (b) and (c) of this section and that the requirements of paragraphs (d) through (l) of this section are complied with.

(b) Lightering operations should be conducted in accordance with the Oil Companies International Marine Forum Ship to Ship Transfer Guide (Petroleum), Second Edition, 1988, to the maximum extent practicable.

(c) Helicopter operations should be conducted in accordance with the International Chamber of Shipping Guide to Helicopter/Ship Operations, Third Edition, 1989, to the maximum extent practicable.

(d) The vessel to be lightered shall make a voice warning prior to the commencement of lightering activities via 13 VHF and 2182 kHz. The voice warning shall include:
    (1) The names of the vessels involved;
    (2) The vessel's geographical positions and general headings;
    (3) A description of the operations;
    (4) The expected time of commencement and duration of the operations; and
    5) Request for wide berth

(l) In preparing to moor alongside the vessel to be lightered, a service vessel shall not approach the vessel to be lightered closer than 1000 meters unless the service vessel is positioned broad on the quarter of the vessel to be lightered. The service vessel must transition to a nearly parallel heading prior to closing to within 50 meters of the vessel to be lightered.

Appendix H

SAFETY CHECKLISTS

OPERATIONAL/SAFETY CHECKLIST

SHIP TO SHIP TRANSFER

CHECK LIST 1 — PRE-FIXTURE INFORMATION

(BETWEEN OPERATORS/CHARTERER(S))

Ship's Name: _____

Operator: _____

Charterer: _____

	Operator's Confirmation	Remarks
1. Is centre of cargo manifold arrangement 3.0 metres or less either forward or aft of mid length position?.	☐	
2. Is centre of cargo manifold at least 0.9 metre above deck, or above working platform if fitted?	☐	
3. Is the height of the centre of cargo manifold no greater than 2.1 metres above the deck?	☐	
4. What is the horizontal spacing between manifold connections, measured centre to centre?	☐	
5. Is ship fitted with a hose support rail at the ship's side constructed of curved plate or piping having a diameter of not less than 200mm?	☐	
6. If a hose support rail is fitted is this at least 700mm below centre if cargo manifold?	☐	
7. Is ship able to present 2 x 400mm manifold connections?	☐	
8. Is ship equipped with sufficient enclosed type fairleads on both sides to receive headlines, sternlines and backsprings from the other ship?	☐	
9. If the answer to question 8 is "yes", are the two fairleads which will receive the other ship's backsprings positioned not more than 35 metres forward and not more than 35 metres aft of the midships position?	☐	
10. Are there bitts of sufficient strength and suitably located inboard of enclosed fairleads to receive eyes of mooring ropes?	☐	
11. Are both sides clear of any overhanging projections?	☐	

For Operator: _____

Position: _____

Signature: _____ Date: _____

FIGURE H-1 Typical lightering safety checklist. Source: OMI Petrolink Corp.

SHIP TO SHIP TRANSFER

CHECK LIST 2 — BEFORE OPERATIONS COMMENCE

Discharging Ship's Name: _____

Receiving Ship's Name: _____

Date of Transfer: _____

	Discharging Ship Checked	Receiving Ship Checked	Remarks
1. Has Check List 1 been completed and ship compatibility established?	☐	☐	
2. Are radio communications established?	☐	☐	
3. Are all walkie-talkie sets in order?	☐	☐	
4. Is language of operation agreed?	☐	☐	
5. Has cathodic protection procedure been checked (see Sections 3.6 and 5.4)?	☐	☐	
6. Has rendezvous position been agreed?	☐	☐	
7. Have method of approach and mooring and unmooring procedures been agreed and decision taken on which ship will provide moorings?	☐	☐	
8. Is ship upright and at suitable trim?	☐	☐	
9. Have engines, steering gear, controls and navigational equipment been tested and found in good order?	☐	☐	
10. Is chief engineer briefed on engine requirements?	☐	☐	
11. Have weather forecasts for transfer area been obtained?	☐	☐	
12. Has hoses lifting equipment been checked and found ready for use?	☐	☐	
13. Are manifold connections ready and marked?	☐	☐	
14. Have hoses been checked and found to be in good order (where applicable)?	☐	☐	
15. Have fenders and handling equipment been checked and found to be in good order (where applicable)?	☐	☐	
16. Is anchor on opposite side to transfer made ready for dropping (where applicable)?	☐	☐	
17. Are navigational signals ready (see Section 5.7)?	☐	☐	
18. Are mooring lines ready both fore and aft?	☐	☐	
19. Are mooring winches in good order?	☐	☐	
20. Are messengers, stoppers and heaving lines in place and ready for use?	☐	☐	
21. Has crew been briefed on mooring methods?	☐	☐	
22. Has a contingency plan been prepared and agreed?	☐	☐	
23. Have authorities been advised (where applicable)?	☐	☐	
24. Has navigational warning been broadcast (where applicable)?	☐	☐	
25. Has other ship been advised that Check List 2 completed in the affirmative?	☐	☐	

FOR DISCHARGING SHIP/RECEIVING SHIP*

Name: _____

Rank: _____

Signature: _____ Date: _____

*Delete as appropriate

SHIP TO SHIP TRANSFER
CHECK LIST 3 — BEFORE RUN-IN AND MOORING

Discharging Ship's Name: _____

Receiving Ship's Name: _____

Date of Transfer: _____

	Discharging Ship Checked	Receiving Ship Checked	Remarks
1. Has Check List 2 been completed?	☐	☐	
2. Are primary fenders floating in place? Have towing and securing lines been checked and found in order? Is handling gear retracted (where applicable)?	☐	☐	
3. Are secondary fenders in place? Is handling gear retracted (where applicable)?	☐	☐	
4. Have any protrusions on outboard or side of berthing been retracted?	☐	☐	
5. Are in-port smoking regulations now in force?	☐	☐	
6. Is proficient helmsman at the wheel?	☐	☐	
7. Are scuppers plugged and sealed?	☐	☐	
8. Has required course and speed information been exchanged and understood?	☐	☐	
9. Are engines controlled by revolutions?	☐	☐	
10. Has area traffic (shipping) been checked?	☐	☐	
11. Are navigational signals displayed (see Section 5.7)?	☐	☐	
12. Are accommodation doors and ports closed?	☐	☐	
13. Is firefighting and anti-pollution equipment checked and ready for use?	☐	☐	
14. Is adequate lighting available, especially overside in vicinity of fenders?	☐	☐	
15. Are hand torches to be used of an approved type?	☐	☐	
16. Have portable transceiver sets been tested and are they intrinsically safe?	☐	☐	
17. Is power on winches and windlass?	☐	☐	
18. Are mooring gangs in position?	☐	☐	
19. Have communications been established with mooring gangs?	☐	☐	
20. Has other ship been advised that Check List 3 completed in the affirmative?	☐	☐	

FOR DISCHARGING SHIP/RECEIVING SHIP*

Name: _____

Rank: _____

Signature: _____ Date: _____

*Delete as appropriate

SHIP TO SHIP TRANSFER

CHECK LIST 4— BEFORE CARGO TRANSFER

Discharging Ship's Name: _____

Receiving Ship's Name: _____

Date of Transfer: _____

	Discharging Ship Checked	Receiving Ship Checked	Remarks
1. Is the gangway in position and secured (where applicable)?	☐	☐	
2. Has communication system been established with other ship?	☐	☐	
3. Have emergency signals and shudown procedures been agreed?	☐	☐	
4. Has bridge watch been established? Has anchor watch been established (where applicable)?	☐	☐	
5. Has efficient deck watch been established with particular attention to mooring, fenders, hoses and manifold observation?	☐	☐	
6. Is there an efficient engineroom watch, and are main engines on standby?	☐	☐	
7. Has initial loading rate been agreed with other ship?	☐	☐	
8. Has maximum loading rate been agreed with other ship?	☐	☐	
9. Has topping-off rate been agreed with other ship?	☐	☐	
10. Are scuppers effectively plugged and drip trays in position under the manifold connections?	☐	☐	
11. Have hoses been tested after connection (where applicable)?	☐	☐	
12. Are hoses supended efficiently?	☐	☐	
13. Are sea and overboard discharge valves of cargo system tightly closed and sealed?	☐	☐	
14. Are tools located at manifold ready for rapid disconnecting?	☐	☐	
15. Are window type air conditioning units (where fitted) disconnected?	☐	☐	
16. Are air conditioning intakes which may permit the entry of cargo vapours closed?	☐	☐	
17. Are fire axes in position fore and aft?	☐	☐	
18. Are all unused manifold connections closed and blanked?	☐	☐	
19. Is firefighting and anti-pollution equipment checked and ready for use?	☐	☐	
20. Is the agreed tank venting system being used?	☐	☐	
21. Is inert gas system Where fitted) operating?	☐	☐	
22. Is radio station closed down and are aerials earthed (grounded) where necessary?	☐	☐	
23. Has other ship been advised that Check List 4 completed in the affirmative?	☐	☐	

FOR DISCHARGING SHIP/RECEIVING SHIP*

Name: _____

Rank: _____

Signature: _____ Date: _____

*Delete as appropriate

SHIP TO SHIP TRANSFER

CHECK LIST 5 — BEFORE UNMOORING

Discharging Ship's Name: _____

Receiving Ship's Name: _____

Date of Transfer: _____

	Discharging Ship Checked	Receiving Ship Checked	Remarks
1. Are cargo hoses or manifold blanked?	☐	☐	
2. Is transfer side of ship clear of obstructions including hose lifting equipment?	☐	☐	
3. Has method of disengagement and of letting go moorings been agreed with other ship?	☐	☐	
4. Have fenders, including towing and securing lines, been checked in good order (where applicable)?	☐	☐	
5. Is power on winches and windlass?	☐	☐	
6. Are messengers, rope stoppers etc., at all mooring stations?	☐	☐	
7. Are crew at stations standing by?	☐	☐	
8. Are communications established with the other ship?	☐	☐	
9. Are communications established with mooring gangs?	☐	☐	
10. Has area traffic (shipping) been checked?	☐	☐	
11. Have mooring crews been instructed to cast off only in the manner and when requested by the maneuvering ship?	☐	☐	
12. Has other ship advised that Check List 5 completed in the affirmative?	☐	☐	
13. Has navigational warning been cancelled when clear of other ship (if applicable)?	☐	☐	

FOR DISCHARGING SHIP/RECEIVING SHIP*

Name: _____

Rank: _____

Signature: _____ Date: _____

*Delete as appropriate

Appendix I

STATEMENT OF TASK

The NRC's Marine Board Committee on Tank Vessel Lightering will conduct a study on oil spill risks from lightering (vessel-to-vessel oil transfer) operations. This study will evaluate current lightering practices and trends and analyze the associated risks. It will make recommendations for appropriate technical and institutional improvements. The study will investigate the incidence and risks of accidents, assess the existing regulatory and management framework and recommend measures that could further reduce oil spill risks. It will take into account the current and proposed international rules and standards.

The spill incidence and risk investigation will consider the vulnerability of lightering operations and the potential for spills under various conditions that are likely to occur. Because some spills associated with lightering operations are caused by human error, these aspects will receive careful examination. The study will also pay particular attention to accident prevention including considerations of operator training, monitoring and inspections.

The committee will first establish the study scope and select the detailed steps to accomplish the tasks. Investigations will be designed to collect data and provide the basis for analyses. Some committee members will visit lightering operations in regions where these practices are conducted to better understand the conditions and the most realistic expectations of any proposed risk reduction measures. The committee will also hold extensive discussions with Coast Guard and other officials who are charged with regulating and managing lightering in local regions. The committee will prepare a final report within twelve months of its first meeting that includes all analyses conducted during the study and presents their final conclusions and recommendations.

The study will be funded by the United States Coast Guard within the Department of Transportation.